Cluster Molecules of the *p*-Block Elements

Catherine E. Housecroft

University Lecturer in Inorganic Chemistry, University of Cambridge

OXFORD   NEW YORK   TOKYO
OXFORD UNIVERSITY PRESS
1994

Oxford University Press, Walton Street, Oxford OX2 6DP

Oxford New York Toronto
Delhi Bombay Calcutta Madras Karachi
Kuala Lumpur Singapore Hong Kong Tokyo
Nairobi Dar es Salaam Cape Town
Melbourne Auckland Madrid

and associated companies in
Berlin Ibadan

Oxford is a trade mark of Oxford University Press

Published in the United States
by Oxford University Press Inc., New York

A catalogue record for this book is available from the British Library

Library of Congress Cataloging in Publication Data

Housecroft, Catherine E., 1955-
Cluster molecules of the p-block elements / Catherine E.
Housecroft. --1st ed.
(Oxford chemistry primers; 14)
Includes index.
1. metal cystals. 2. Chemical elements. I. Title. II. Title:
p-block elements. III. Series.
QD171.H83 1983   546'.3 --dc20

ISBN 0 19 855699 3 (Hbk)
ISBN 0 19 855698 5 (Pbk)

Typeset by the author
Printed in Great Britain by The Bath Press, Bath, Avon

Series Editor's Foreword

Cluster compounds of the main groups have challenged the imagination of chemists. Many of them are so-called electron deficient compounds that have spurred improvements in bonding theory. The structures themselves often have elegance and symmetry. With the advent of fullerenes, this field will continue to entice and remain a key feature of undergraduate chemistry courses.

Oxford Chemistry Primers are designed to give a concise introduction to all chemistry students by providing the material that would usually be covered in an 8–10 lecture course. As well as providing up-to-date information, this series will provide explanations and rationales that form the framework of an understanding of inorganic chemistry. Catherine Housecroft has provided us with a mine of chemical information woven into the present-day understanding of these fascinating compounds that should be appreciated by all undergraduates.

John Evans
Department of Chemistry, University of Southampton

Preface

The *p*-block elements make up the right hand side of the periodic table. *Cluster molecules of the p-block elements* deals with cluster molecules formed by the elements of this block—this is an area of research that has seen tremendous growth in the last decade and it is continually growing. In the elemental state, elements like phosphorus and carbon form discrete clusters such as P_4 and C_{60}. In the solid state, allotropes of boron form lattices in which icosahedral cluster units are the central building blocks. In their compounds, many *p*-block elements, in particular those from groups 13–16, show a tendency to form cluster molecules, for example, boranes, cubanes, adamantane-like clusters, Zintl ions and small clusters containing carbon atoms. The bonding in many cluster molecules presents a challenge to the theorist—boranes, for example, are termed electron deficient since there are fewer valence electrons available than contacts between neighbouring pairs of atoms. This book provides an introduction to a large and diverse area of chemistry. It is arranged so as to provide the reader with separate discussions of the elemental state, structure, bonding, synthesis, and reactivity. A list of sources of further reading in Chapter 1 should enable the reader to probe the current literature.

My thanks go to my husband, Edwin Constable, who has been a constant source of encouragement and help. No writing was ever possible without the presence of Philby and Isis—up to Siamese tricks or deep in feline slumbers!

Cambridge
December 1992

C.E.H.

Contents

Series sponsor: **ZENECA**

ZENECA is a major international company active in four main areas of business: Pharmaceuticals, Agrochemicals and Seeds, Specialty Chemicals, and Biological Products.

ZENECA's skill and innovative ideas in organic chemistry and bioscience create products and services which improve the world's health, nutrition, environment, and quality of life.

ZENECA is committed to the support of education in chemistry.

1 Introduction and definitions

1.1 The *p*-block elements

The *p*-block elements are those lying in groups 13 to 18 of the periodic table. This text focuses attention on elements in groups 13 to 16. Each group 13 element possesses three valence electrons, each group 14 element, four, each group 15 element, five, and each group 16 element, six. These numbers are fundamental to an understanding of the chemistry of the *p*-block elements.

1.2 What is a cluster?

In this book, a *cluster* is defined as a neutral or charged species in which there is a polycyclic array of atoms (Fig. 1.1). The cluster may be homo- or hetero-atomic. It may be naked with no directly bonded peripheral atoms as in elemental P_4 or the Zintl ion $[Pb_5]^{2-}$. In other cases, the cluster may exhibit a central core. For example, $[B_6H_6]^{2-}$ consists of a central B_6-unit in which a terminal hydrogen atom is attached to each boron atom. Edges of the cluster-core may be bridged by atoms (e.g. H) without total loss of a bonding interaction between the cluster atoms (e.g. B_6H_{10}). These differences may be rationalized in terms of bonding and will be discussed in later chapters. Many transition metal cluster compounds are known (e.g. $Rh_4(CO)_{12}$, $Os_6(CO)_{18}$, and $[Fe_4(CO)_{13}]^{2-}$) but these are beyond the scope of this book.

Electron deficient: This is a term that is used when the number of available valence electrons is inadequate to provide a 2-centre-2-electron bonding scheme.

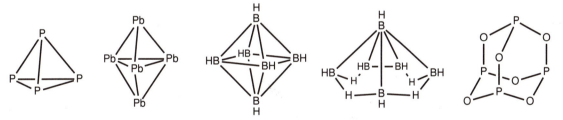

Fig. 1.1 Examples of clusters formed by *p*-block elements; structures of P_4, $[Pb_5]^{2-}$, $[B_6H_6]^{2-}$, B_6H_{10}, and P_4O_6.

Bonding descriptions for clusters vary depending upon the availability of valence electrons; methods of rationalizing the bonding in cluster molecules will be described in Chapter 4. Clusters involving boron atoms are often described as being *electron deficient*; such molecules possess an inadequate number of valence electrons to be held together by a network of localized 2-centre-2-electron bonds. On the other hand, a P_4 tetrahedron has enough valence electrons to allocate a localized 2-centre-2-electron bond to each edge and there is still a lone pair left over per phosphorus atom. It is important therefore to recognize that a line drawn between two atoms in the

A line drawn between two atoms in a cluster does not necessarily imply the presence of a localized bond.

A *deltahedron* is a specific type of polyhedron—one with all triangular faces.

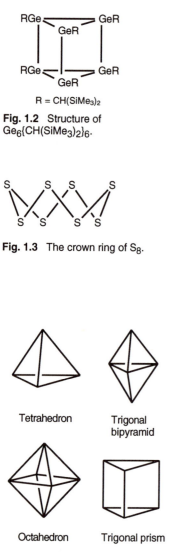

R = CH(SiMe₃)₂

Fig. 1.2 Structure of Ge₆{CH(SiMe₃)₂}₆.

Fig. 1.3 The crown ring of S₈.

Tetrahedron   Trigonal bipyramid

Octahedron   Trigonal prism

Fig. 1.4 Polyhedral cages for $n = 4$ to 6.

cluster framework does not necessarily imply the presence of a localized bond; in P_4 and P_4O_6 it does, but in $[Pb_5]^{2-}$, $[B_6H_6]^{2-}$, and B_6H_{10} it does not.

Some clusters are more open than others—compare P_4 and P_4O_6 in Fig. 1.1. The terms *closed* and *open* clusters are used commonly but a definition of these terms is not trivial because electron deficient clusters cannot be treated in the same way as other species. For an electron deficient cluster, a closed cage corresponds to the parent *deltahedron*, e.g. the cage structure of $[B_6H_6]^{2-}$ (Fig. 1.1) is an octahedron and is closed. The surface of the cage shows no deviation from triangular faces. If a departure from triangular faces is observed, e.g. in B_6H_{10} (Fig. 1.1), then the electron deficient cluster is described as being open. This concept is examined in Chapter 4.

Cluster molecules in which there is an adequate number of electrons to form localized 2-centre-2-electron bonds between adjacent atoms are not restricted to deltahedral structures but exhibit instead a wide range of polycyclic frameworks. $Ge_6\{CH(SiMe_3)_2\}_6$ (Fig. 1.2) adopts a trigonal prismatic structure and not an octahedral one. The trigonal prism is more open than the octahedron but for a species in which there are plenty of bonding electrons available, it is a natural choice (see Chapter 4) and describing $Ge_6\{CH(SiMe_3)_2\}_6$ as an open cluster could be misleading.

At this stage, the reader should be aware that there is a descriptive problem. There comes a point when the threshold between a *cluster* and a *ring* is crossed. In elemental sulfur (Fig. 1.3), the eight atoms are arranged in a single ring. Could such a ring be classed as an open cluster? For S_8, this seems to be an unnecessary complication.

In this book, the term *ring* will be reserved for monocyclic and bicyclic molecules. A *cluster* will be defined as a molecule possessing at least a tricyclic structure. An exception is the borane B_4H_{10} (Fig. 3.1). There is one large group of compounds that cannot fall into this definition—in general, organic molecules containing fused rings cannot realistically be classed as clusters although some possible candidates are discussed in Section 3.5.

1.3 Polyhedral cages

The structures of many of the clusters described in this book are based upon regular polyhedra. For electron-deficient clusters, these parent polyhedra exhibit triangular faces only, i.e. they are *deltahedra*. The term *parent polyhedron* is used because many clusters possess incomplete polyhedral frameworks and such structures are most readily described in terms of being derived from a complete polyhedron by the removal of one or more vertices. This concept will be addressed in Chapter 4.

The minimum number of atoms that can form a closed (three dimensional) cluster molecule is four. Throughout this book, the number of cluster vertices will be defined as n. For $n = 4$, the tetrahedron is the only closed polyhedron which may be adopted (Fig. 1.4). For $n = 5$, the trigonal bipyramid is usually observed (Fig. 1.4). For $n = 6$, atoms may be arranged in two triangles that are either mutually eclipsed or staggered to give the trigonal prism or octahedron, respectively (Figs. 1.4 and 1.5).

Some commonly encountered polyhedral cages are shown on the front inside cover of this book. The pentagonal bipyramid ($n = 7$), dodecahedron ($n = 8$), hexagonal bipyramid ($n = 8$), tricapped trigonal prism ($n = 9$), bicapped antiprism ($n = 10$), octadecahedron ($n = 11$), and icosahedron ($n = 12$) are all deltahedra. For $n = 8$, one structural possibility is the simple cube. The square antiprism (related to the cube by rotating one pair of opposing square faces so that they are mutually staggered) is a structure that must also be considered for $n = 8$ since the energy barrier between the dodecahedron, the square antiprism and the hexagonal bipyramid is low. For $n = 9$, a second polyhedral option is the monocapped antiprism. This is related to the tricapped trigonal prism by breaking the bottom edge of the central prism of the tricapped trigonal prism and allowing the resultant 4-sided face to open out into a square. This generates the monocapped square antiprism. For $n = 12$, the cubeoctahedron and anticubeoctahedron are accessible structures in addition to the icosahedron.

The polyhedra shown in Fig. 1.4 and on the inside book-cover constitute a restricted set, restricted because polyhedra with faces larger than a square are not included. Examples of closed clusters of the *p*-block elements exhibiting pentagonal or hexagonal faces are not common but examples that are known are important: elemental boron and C_{60} will be discussed in Chapter 2.

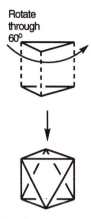

Fig. 1.5 Conversion of a trigonal prism to octahedron.

1.4   Further reading

Wells, A.F. (1984). *Structural Inorganic Chemistry*, 5th Edn, OUP, Oxford.

Kroto, H.W. (1992). C_{60}: Buckminsterfullerene, The Celestial Sphere that Fell to Earth, *Angewandte Chemie, Int. Ed., Engl.*, **31**, 111–129.

Corbridge, D.E.C. (1974). *The Structural Chemistry of Phosphorus*, Elsevier, Amsterdam.

Baudler, M. (1987). Polyphosphorus Compounds—New Results and Perspectives, *Angewandte Chemie, Int. Ed., Engl.*, **26**, 419–441.

Greenwood, N.N. and Earnshaw, A. (1985). *Chemistry of the Elements*, Pergamon Press, Oxford.

Greenwood, N.N. (1989). Boron Hydride Compounds. Chapter 2 in *Rings, Clusters and Polymers of Main Group and Transition Elements* (ed. H.W. Roesky), Elsevier, Amsterdam.

Housecroft, C.E. (1990). *Boranes and Metalloboranes: Structure, Bonding and Reactivity*, Ellis Horwood, Chichester.

Kennedy, J.D. (1984, 1986). The Polyhedral Metallaboranes, Parts I and II. *Progress in Inorganic Chemistry*, **32**, 519–679; **34**, 211–434.

Fort, R.C. and Schleyer, P. von R. (1964). Adamantanes: consequences of the Diamondoid Structure. *Chemical Reviews*, **64**, 277–300.

Haiduic, I. and Sowerby, D.B. (1987). *The Chemistry of Inorganic Homo- and Heterocycles*, Vols. 1 and 2, Academic Press, New York.

Wade, K. (1976). Structural and Bonding Patterns in Cluster Chemistry. *Advances in Inorganic Chemistry and Radiochemistry*, **18**, 1–66.

Mingos, D.M.P. and Wales, D.J. (1990). *Introduction to Cluster Chemistry*, Prentice Hall, New Jersey.

Chapter 2
Structures of the elements
The story of C_{60}

Phosphorus

Chapters 3-6
Structure, bonding, synthesis, and reactivity

2 Clusters in the elemental state

2.1 Which elements form discrete clusters in their elemental states?

In this chapter, the occurrence of clusters for members of the *p*-block in their elemental states is described. Distinctions are made between elements which do not form clusters at all, elements for which cluster units form part of an extended lattice, and elements which exhibit discrete (i.e. separated) clusters in at least one allotrope.

All the inert (noble) gases (group 18) are monatomic; this is a consequence of the complete electronic configuration exhibited by each element. The members of group 17, the halogens, are all diatomic molecules. The general ground state valence configuration of a halogen atom X is $ns^2 np^5$ and therefore a single bond to another halogen atom to form X_2 provides an octet of valence electrons around each atom, i.e. the *octet rule* is satisfied (Eqn 2.1).

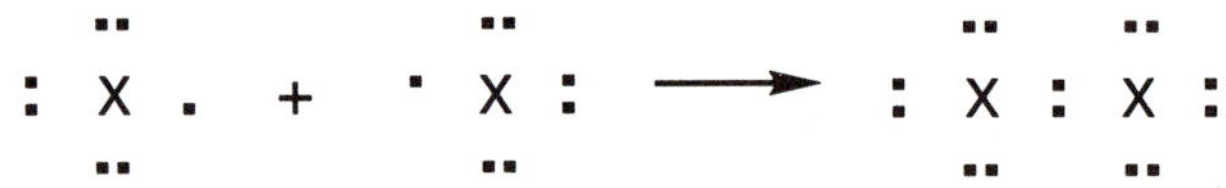

The Octet Rule: An *s*- or *p*-block element obeys the *octet rule* if the nucleus is surrounded by eight valence electrons.
e.g. In NH_3, the formation of three N–H bonds introduces three electrons into the valence shell of the N atom and the N nucleus then has an octet of valence electrons.

$$:\overset{\displaystyle ..}{\underset{\displaystyle ..}{X}}\cdot \;+\; \cdot \overset{\displaystyle ..}{\underset{\displaystyle ..}{X}}: \;\longrightarrow\; :\overset{\displaystyle ..}{\underset{\displaystyle ..}{X}}:\overset{\displaystyle ..}{\underset{\displaystyle ..}{X}}: \qquad \textbf{Eqn 2.1}$$

In group 16, the ground state valence electronic configuration of an element E is $ns^2 np^4$. This permits the formation of two covalent bonds. In compounds of sulfur and heavier elements in group 16, an expansion of the valency to four or six may be possible by the use of low lying *d*-orbitals if they are available. There are three ways in which an atom E may satisfy the octet rule in the elemental state: (i) by the formation of a diatomic molecule containing an E=E double bond (Figs. 2.1 and 2.2), (ii) by the formation of a cyclic molecule (Fig. 2.2), or (iii) by the formation of a polymeric chain of atoms (Fig. 2.2). In the elemental state at standard temperature and pressure, oxygen alone in the group favours diatomic molecule formation. The triatomic molecule, ozone, (Fig. 2.1), is thermodynamically unstable with respect to O_2 (Eqn 2.2). The experimental O–O distances in O_3 indicate some degree of interaction between the terminal oxygen atoms of the V-shaped molecule.

O=O
1.211 Å

O — 1.278 Å

2.18 Å

Fig. 2.1 O_2 and O_3.

$$O_3\,(g) \;\longrightarrow\; {}^3\!/_2\,O_2\,(g) \qquad \Delta G^o = -163.2 \text{ kJ mol}^{-1} \qquad \textbf{Eqn 2.2}$$

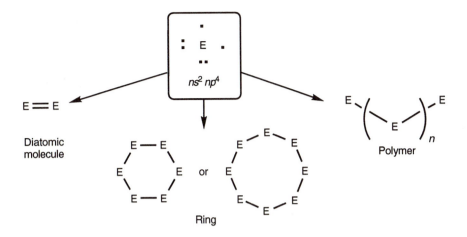

Fig. 2.2 Molecule formation in the elemental state for a group 16 element, E. Rings could theoretically be of any size.

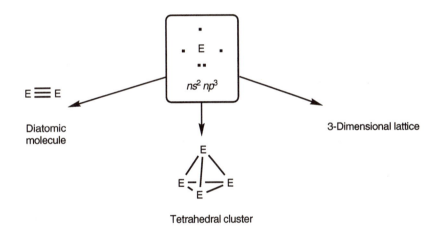

Fig. 2.3 Structural choices in the elemental state for a group 15 element, E.

In group 15, the ground state valence electronic configuration of $ns^2 np^3$ allows an atom to form three covalent bonds. In compounds of phosphorus and its heavier congeners, participation of low lying *d*-orbitals may allow an expansion of the valency from three to five. In its native state, an atom of the first element of group 15, nitrogen, satisfies the octet rule by forming a triple bond to another nitrogen atom. The N≡N bond is very strong (945 kJ mol⁻¹) and the formation of diatomic N_2 is very favourable. For elements lower in the group, the formation of a tetrahedral cluster in the elemental state is observed, although the molecule E_4 occurs with varying degrees of stability depending on E (see Sections 2.3 and 2.4). Three-dimensional lattices are the alternative possibility; for example, a network of puckered six-membered rings is observed for black phosphorus and related structures are exhibited by *α*-arsenic, *α*-antimony, and *α*-bismuth. The reader is directed to references given in Section 1.4 for further details of three-dimensional arrays.

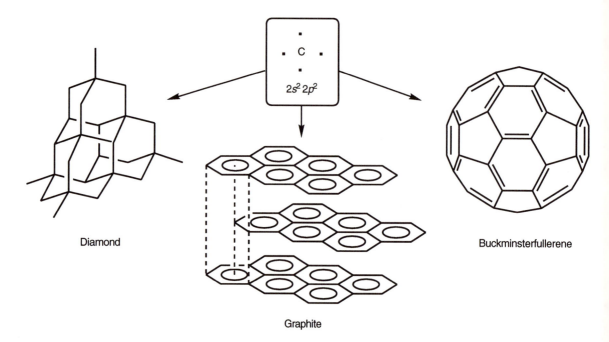

Fig. 2.4 Structural diversity in the elemental state of carbon. The carbon atoms in each layer of graphite are involved in π-bonding giving π-delocalization throughout the layer. The layers of fused hexagons in graphite are staggered and alternate in an *a-b-a* pattern. More detailed diagrams of buckminsterfullerene are given in Fig. 2.7.

The characteristic ground state valence electronic configuration of a group 14 element is $ns^2 np^2$. In the absence of *d*-orbital participation four covalent bonds may be formed per atom. The allotropic forms of carbon are diamond, graphite, and the recently discovered fullerenes (Fig. 2.4). Of these, only the fullerenes are discrete clusters and will be discussed in Section 2.2; C_{60} (buckminsterfullerene) is shown schematically in Fig. 2.4. Silicon and germanium both crystallize with a diamond-type lattice. Tin and lead are metallic. Solid lead possesses a close-packed structure. The low temperature form of tin (grey or α-tin) exhibits the diamond structure but at 13.2°C lattice distortion occurs to form white β-tin in which each tin atom has six nearest neighbours.

In group 13, the ground state valence configuration of $ns^2 ns^1$ would permit the formation of three covalent bonds but such implied simplicity of structure is not observed in the elemental state. The metallic character of aluminium and its heavier congeners leads to close-packed lattices for these elements. The atoms of elemental boron tend to aggregate in an effort to overcome the problem of each atom possessing an empty $2p$ atomic orbital were it only to form three covalent bonds. This phenomenon is also seen at a molecular level in its compounds (see Chapters 3 and 4). In the elemental state, the clusters which form as a result of atomic aggregation are *not* discrete although the icosahedral B_{12}-cluster (Fig. 2.5) is a recurrent structural building block.

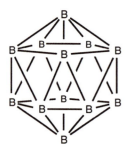

Fig. 2.5 An icosahedral B_{12}-cluster which is a recurrent structural unit in allotropes of boron.

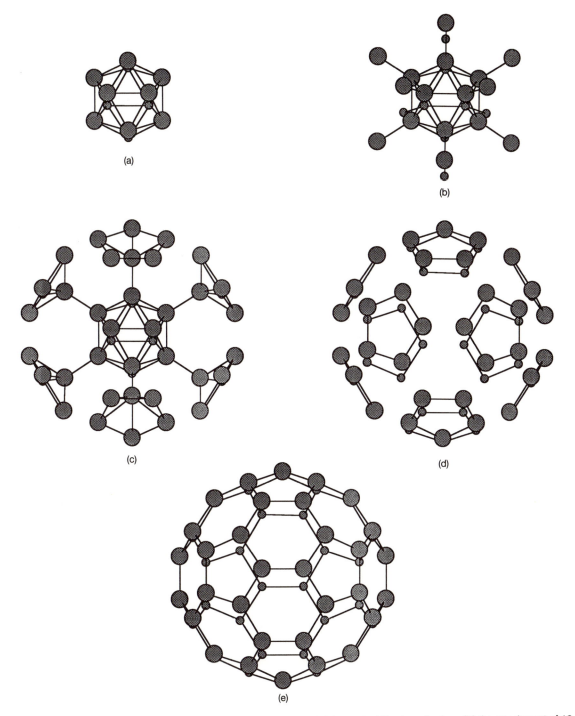

Fig. 2.6 The construction of the B_{84}-unit in β-rhombohedral boron: (a) the central B_{12}-icosahedron, (b) the attachment of 12 terminal boron atoms to the central B_{12}-icosahedron, (c) the incorporation of each terminal boron atom into a pentagonal pyramidal unit (only 6 of the 12 are shown for clarity), (d) the network of 12 isolated pentagons arising from the 12 pentagonal pyramids, (e) the B_{60}-cage formed by joining together the 12 pentagons. Note the direct relationship between the outer B_{60}-cage and buckminsterfullerene, C_{60}.

The first polymorph of boron to be described was the α-tetragonal form but this has been reformulated as a carbide or nitride, $B_{50}C_2$ or $B_{50}N_2$. The carbon or nitrogen content is present as a result of synthetic conditions. In the α-rhombohedral allotrope, B_{12}-clusters are arranged in an approximately close-packed lattice. Boron–boron interactions between the icosahedral clusters are weaker than those within a B_{12}-unit. The structure of β-rhombohedral boron is complex and consists of B_{84}-units connected by B_{10}-units and single boron atoms. There are 105 atoms in the unit cell and the formula B_{105} may be usefully written as $(B_{84})(B_{10}.B.B_{10})$. The B_{84}-cage is best considered by building it up from the icosahedral cluster which lies at its centre (Fig. 2.6a). A terminal boron atom is bonded to each atom of the B_{12}-cluster (Fig. 2.6b). Each of the 12 terminal atoms is incorporated (as the apical atom) into a pentagonal pyramid of boron atoms. Six of these pentagonal pyramids are shown in Fig. 2.6c. There are twelve pentagonal pyramids in the complete 3-dimensional cage and the network of twelve pentagons thereby produced is shown in Fig. 2.6d. When joined together, these pentagons form an approximately spherical B_{60}-cage (Fig. 2.6e) identical to that of C_{60} (Figs. 2.4 and 2.7). Figure 2.6 illustrates that the B_{84}-cage of β-rhombohedral boron may be represented as $(B_{12})(B_{12})(B_{60})$.

After considering the elements of the *p*-block, it is clear that only a few elements exhibit discrete cluster units in their native state. These are detailed below.

2.2 Fullerenes

The characterization of C_{60} and C_{70}

The name *fullerene* has been coined after the architect Buckminster Fuller who is well known for his geodesic domes.

The emergence of the chemistry of the carbon cluster C_{60}, buckminster-fullerene, has occurred since 1985, and dramatically so since a method for producing bulk quantities of C_{60} and C_{70} was developed in 1990.

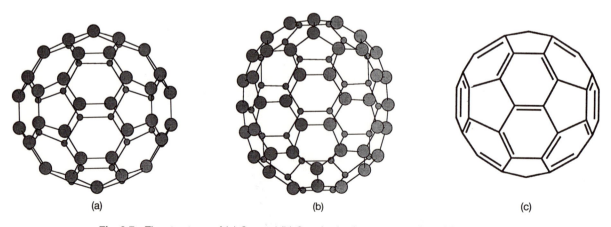

(a)   (b)   (c)

Fig. 2.7 The structures of (a) C_{60} and (b) C_{70}. A simpler representation of C_{60} is shown in (c).

Pure graphitic soot (which contains a few per cent by weight of molecular C_{60}) is produced by evaporating graphite rods in an atmosphere of helium at a pressure of about 100 torr and condensing the vapour. Upon dispersal in benzene, some of the condensed product dissolves to give a red solution from which a black crystalline material is obtained. The dominant components of this material are C_{60} and C_{70} (Fig. 2.7); a mass spectrum shows a strong peak at 720 a.m.u. and a less intense peak at 840 a.m.u. corresponding to C_{60} and C_{70} respectively. Evidence is available for higher fullerenes such as C_{84} but data concerning these molecles are, as yet, sparse.

Magenta C_{60} and red C_{70} may be separated chromatographically by using hexane. The initial characterization of C_{60} and C_{70} relied heavily upon spectroscopic methods. For C_{60}, the infrared spectrum exhibits four absorptions at 1429, 1183, 577, and 527 cm^{-1}; this pattern is in accordance with the number of vibrational modes predicted for a cage with icosahedral (I_h) symmetry. Raman spectroscopic data are consistent with this result. In 1990, X-ray diffraction studies on platelets of C_{60} (i.e. not a single crystal diffraction study) indicated that the solid consisted of an ordered array of spheroidal molecules, each ≈7 Å in diameter and separated from its neighbour by ≈3 Å. The ^{13}C NMR spectrum of C_{60} shows a single resonance at δ 143 indicating that all 60 carbon atoms are in equivalent sites; the chemical shift compares quite well with that observed for benzene at δ 128. The results of an electron diffraction study of gaseous C_{60} (in 1991) provided evidence for the structure shown in Fig. 2.7a. All atoms are equivalent but there are two distinct types of bond: those between hexagonal faces (a C_6–C_6 bond) and those at the junction of one hexagonal and one pentagonal face (a C_6–C_5 bond). Distances in the gas phase are 1.40 Å and 1.46 Å respectively and may be designated as double C=C and single C–C bonds. In mid-1992, the single crystal structure of C_{60} was reported and confirms the structure shown in Fig. 2.7a; interatomic distances are about 0.01 Å shorter than those obtained from the gas phase study.

An important feature of the C_{60} and C_{70} cluster frameworks is that each is constructed from hexagonal and pentagonal faces; the pentagons are isolated from one another.

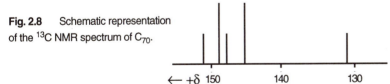

Fig. 2.8 Schematic representation of the ^{13}C NMR spectrum of C_{70}.

The structure of C_{70} has yet to be determined by a single-crystal X-ray diffraction study although the cage shown in Fig. 2.7b is consistent with spectroscopic data and is the same as the cage crystallographically confirmed in the derivative (η^2-C_{70})Ir(CO)Cl(PPh$_3$)$_2$. The molecular structure of C_{70} may be derived from that of C_{60} by splitting the latter into two hemispherical-C_{30} units and rejoining them after the insertion of a band of ten carbon atoms. A molecule of C_{70} possesses D_{5h} symmetry and the appearance of five resonances (intensity ratio 1:2:1:2:1) in the ^{13}C NMR spectrum (Fig. 2.8) indicates that the molecule is not fluxional in benzene solution on the NMR spectroscopic timescale. Two-dimensional ^{13}C–^{13}C NMR spectroscopy permits the correlation of a ^{13}C NMR spectral signal due to one nucleus with one or more signals arising from one or more carbon nuclei to which the first atom is bonded. In this way, the connectivities of the atoms may be deduced and results are in agreement with the structure shown in Fig. 2.7b.

The standard enthalpy of combustion of C_{60} has been determined by bomb calorimetry to be $-25,890.8$ kJ mol^{-1} and this gives a value of ΔH_f°(cryst) $= +2,280.2 \pm 5.6$ kJ mol^{-1}.

The reactivity of C_{60} and C_{70}

Of C_{60} and C_{70}, the former has received the greater attention. The atoms in the C_{60} cluster are regarded as being bonded together by double and single carbon–carbon bonds. C_{60} is highly reactive towards radical species and has been termed a radical-sponge. In the reaction between C_{60} and PhCH$_2^{\bullet}$, up to 15 radicals add to the cluster. Alkyl radicals, R$^{\bullet}$, add rapidly to C_{60}. For R = ^{t}Bu, the ESR spectroscopic signal assigned to ^{t}BuC$_{60}^{\bullet}$ increases as the temperature of a benzene solution of the species in raised from 300 K to 350 K. The process is reversible and these data suggest that the radical formed is in equilibrium with a dimeric species (Eqn 2.3). The structure proposed for the dimer $\{^t$BuC$_{60}\}_2$ is given in Fig. 2.9. The stability of $\{$RC$_{60}\}_2$ depends upon the steric requirements of R.

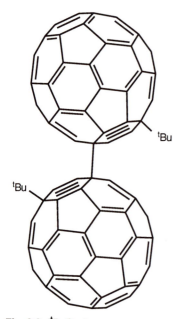

$$2\ ^t\text{BuC}_{60}{}^{\bullet} \rightleftharpoons {}^t\text{BuC}_{60}\text{--C}_{60}{}^t\text{Bu} \qquad \textbf{Eqn 2.3}$$

The reactivity of a C_{60} molecule does not reflect that of benzene even though the surface of the cluster is reminiscent of connected benzene rings. C_{60} reacts with nucleophiles and the polyfunctional nature of the C_{60} cluster is a source of non-selective product formation. One method of controlling the number of ^{t}butyl or ethyl substituents introduced into the cluster is to titrate ^{t}BuLi or EtMgBr against C_{60}. After protonation, the monoalkylated derivatives C_{60}H^{t}Bu and C_{60}HEt are produced, respectively. Further reaction can occur to give more highly substituted products.

C_{60} undergoes Birch reduction (Eqn 2.4) to give a mixture of isomers of $C_{60}H_{36}$. Reoxidation with the quinone shown in Eqn 2.4 regenerates the C_{60} cluster. The reduction potentials for the stepwise reduction of C_{60} and C_{70} are given in Eqn 2.5. Significantly, the first and second reduction steps for the two clusters are very similar but it is easier to reduce $[C_{70}]^{2-}$ than $[C_{60}]^{2-}$ to the corresponding trianion.

Fig. 2.9 $\{^t$BuC$_{60}\}_2$.

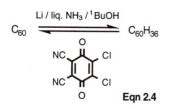

Eqn 2.4

The reduction potentials in Eqn 2.5 are measured with respect to a standard calomel electrode.

$$C_{60} \underset{}{\overset{-612\ \text{mV}}{\rightleftharpoons}} [C_{60}]^- \underset{}{\overset{-1000\ \text{mV}}{\rightleftharpoons}} [C_{60}]^{2-} \underset{}{\overset{-1482\ \text{mV}}{\rightleftharpoons}} [C_{60}]^{3-}$$

Eqn 2.5

$$C_{70} \underset{}{\overset{-616\ \text{mV}}{\rightleftharpoons}} [C_{70}]^- \underset{}{\overset{-988\ \text{mV}}{\rightleftharpoons}} [C_{70}]^{2-} \underset{}{\overset{-1404\ \text{mV}}{\rightleftharpoons}} [C_{70}]^{3-}$$

Expansion (or the so-called inflation) of the C_{60} cluster occurs during reaction with diphenyldiazomethane (Fig. 2.10). The structure of the product has been confirmed in an X-ray diffraction study of $C_{61}(C_6H_4$-4-Br$)_2$. The carbon–carbon distance at the point in the fullerene at which the diphenylmethylene fragment has added is 1.84 Å. This indicates that expansion from a C_{60} to C_{61} cage has indeed occurred.

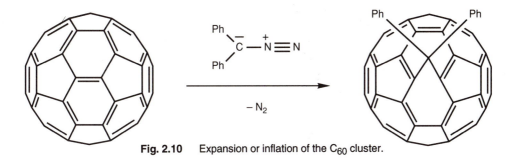

Fig. 2.10   Expansion or inflation of the C_{60} cluster.

The preferential reactivity of a C_6–C_6 bond in C_{60} has been observed in several reactions with transition metal complexes (Fig. 2.11). Recall that in C_{60}, double bond character is associated with the C_6–C_6 edges. C_{60} displaces ethene from $(Ph_3P)_2Pt(\eta^2\text{-}C_2H_4)$. Related to $(Ph_3P)_2Pt(\eta^2\text{-}C_{60})$ is $\{(Et_3P)_2Pt\}_6(C_{60})$ in which six PtL_2 (L = phosphine) units are bound to six separate C_6–C_6 junctions. Both C_{60} and C_{70} exhibit alkene-like character in addition reactions with $(Ph_3P)_2Ir(CO)Cl$ to give $(\eta^2\text{-}C_{60})Ir(CO)Cl(PPh_3)_2$ and $(\eta^2\text{-}C_{70})Ir(CO)Cl(PPh_3)_2$ respectively. In each product, the two carbon atoms that are directly bonded to the transition metal atom are pulled out from the fullerene cluster towards the metal atom.

Modes of attachment of an alkene to a metal atom.

$$\begin{matrix} H_2 \\ C \\ \| \\ C \\ H_2 \end{matrix} \rightarrow ML_n \quad \leftrightarrow \quad \begin{matrix} H_2 \\ C \\ | \\ C \\ H_2 \end{matrix} \!\!\! \rangle ML_n$$

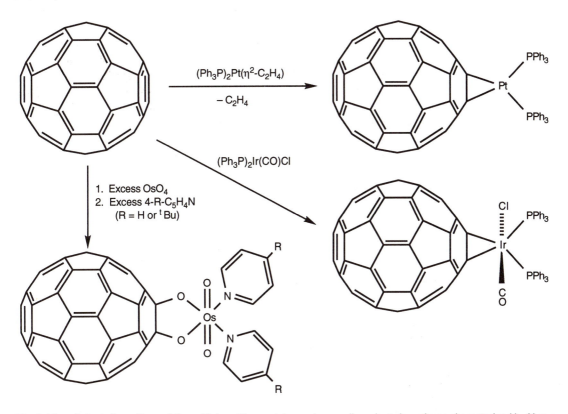

Fig. 2.11   Selected reactions of C_{60} with transition metal complexes; all products have been characterized by X-ray crystallography.

The osmylation of C_{60} in the presence of a pyridine base (Fig. 2.11) gives an adduct $C_{60}O_2Os(O)_2(NC_5H_4\text{-}4\text{-}R)_2$ (R = H or ^{t}butyl). In the case of R = ^{t}Bu, the loss of π-character from the C_6–C_6 bond which interacts with the osmium fragment is confirmed by an increase in carbon–carbon bond length from 1.45 Å in C_{60} to 1.62 Å in $C_{60}O_2Os(O)_2(NC_5H_4{}^tBu)_2$.

The cavity within a fullerene cluster may be occupied by a metal atom of an appropriate size. The laser vaporization of graphite impregnated with La_2O_3 leads to LaC_{60} and ESR spectroscopic data indicate that the compound should be formulated as $[La^{3+}][C_{60}{}^{3-}]$. A fullerene with a metal atom, M, within the cluster is called an *endohedral metallafullerene* and is designated as $M@C_x$. Members of this series include fullerene hosts other than C_{60}, e.g. $Sc_2@C_{82}$, $Sc_3@C_{82}$, $U@C_{28}$, $K@C_{44}$, and $Cs@C_{48}$.

Doping C_{60} with potassium or rubidium produces K_3C_{60} or Rb_3C_{60}. These species are metallic at room temperature but become superconducting at –255°C and –245°C respectively. By comparison with other superconducting materials, alkali metal doped C_{60} exhibits a high onset temperature of superconductivity and this property has evoked tremendous interest in fullerene chemistry as a whole.

2.3   White phosphorus, P₄

Structural and thermodynamic considerations

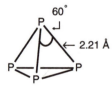

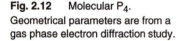

Fig. 2.12   Molecular P_4. Geometrical parameters are from a gas phase electron diffraction study.

White phosphorus is a metastable state and forms upon condensation of phosphorus vapour. Crystalline white phosphorus exhibits two structural modifications. Below –77°C, the hexagonal β-form is present and above this temperature, the cubic α-form exists. Both the α- and β-forms consist of discrete P_4 molecules (Fig. 2.12) which persist when white phosphorus melts (–228.9°C) and vaporizes (7.5°C). Above 554°C, P_4 dissociates into diatomic P_2; the dissociation energy for the process $P_4 \rightarrow 2\,P_2$ is 217 kJ mol^{-1}. The dissolution of white phosphorus in non-aqueous solvents such as CS_2, PCl_3, or liquid ammonia also occurs with retention of the P_4 molecules. P_4 is insoluble in water; it is stored in water to prevent oxidation (Fig. 2.13).

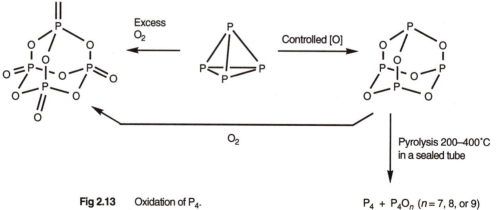

Fig 2.13   Oxidation of P_4.

$P_4 + P_4O_n$ (n = 7, 8, or 9)

Reactivity of P_4

White phosphorus is soft and inflammable. In air, it spontaneously oxidizes to P_4O_{10} (commonly referred to as phosphorus pentoxide, "P_2O_5") although with a limited supply of oxygen P_4O_6 is formed. At high temperatures P_4O_6 disproportionates to P_4 and higher oxides (Fig. 2.13). White phosphorus does not react with water, but does react with aqueous alkali to yield the hypophosphite anion (Eqn 2.6).

The structures of phosphorus oxides and related compounds are discussed in Section 3.8. See Section 4.3 for a discussion of the bonding in P_4, P_4O_6, and P_4O_{10}.

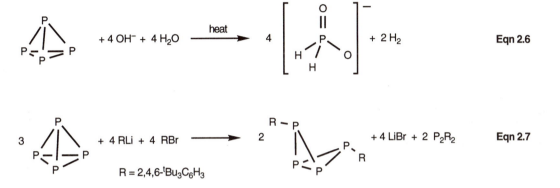

Eqn 2.6

Eqn 2.7

$R = 2,4,6\text{-}^tBu_3C_6H_3$

The P_4 tetrahedron opens up into a *butterfly* conformation when it is effectively reduced according to Eqn 2.7. The internal dihedral angle of the P_4 butterfly is 95.5° and the substituents are arranged in an *exo,exo*-configuration.

In a butterfly or similar cluster, *exo*- and *endo*-substituents are defined as follows:

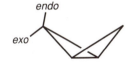

$$23\,P_4 + 12\,LiPH_2 \longrightarrow 6\,Li_2P_{16} + 8\,PH_3 \qquad \text{Eqn 2.8}$$

The compound Li_2P_{16} is formed when white phosphorus is treated with $LiPH_2$ (Eqn 2.8); Li_3P_7 is also produced. A change in the ratio of reactants in Eqn 2.8 may lead to the formation of Li_3P_{21} and Li_4P_{26} in place of Li_2P_{16}. Like $[P_{16}]^{2-}$, (Fig. 2.14), the anions $[P_{21}]^{3-}$ and $[P_{26}]^{4-}$ have structures which are related to an allotrope of the element, Hittorf's (monoclinic) phosphorus (Fig. 2.15). ^{31}P NMR spectroscopic data suggest that discrete $[P_{16}]^{2-}$, $[P_{21}]^{3-}$, and $[P_{26}]^{4-}$ anions exist in solutions of Li_2P_{16}, Li_3P_{21}, and Li_4P_{26}, respectively. Each 2-coordinate phosphorus atom formally bears one negative charge.

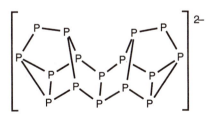

Fig. 2.14 Structure of the dianion $[P_{16}]^{2-}$.

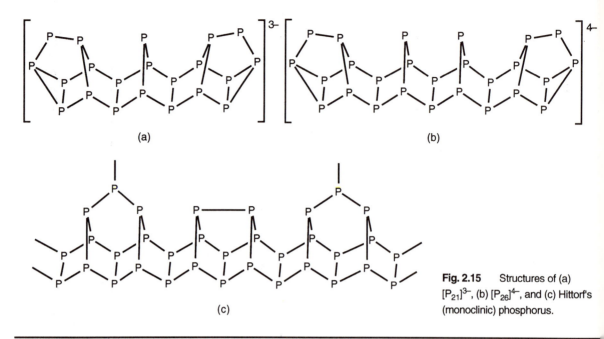

Fig. 2.15 Structures of (a) $[P_{21}]^{3-}$, (b) $[P_{26}]^{4-}$, and (c) Hittorf's (monoclinic) phosphorus.

The eighteen electron rule

The eighteen electron rule is an extension of the octet rule as applied to transition metals. In a low oxidation state, a transition metal will accept electrons from ligands so that the number of valence electrons (ve) surrounding the metal is 18.

e.g. Ni(0) has a ground state electronic configuration of $4s^2\,3d^8$ (10 ve) and requires eight more electrons from surrounding ligands. $\therefore$ It forms compounds such as $Ni(CO)_4$ in which there are four 2-electron donor ligands.

Since each phosphorus atom in P_4 carries a lone pair of electrons, the cluster can function as a Lewis base. In Fig. 2.16a, the coordination of one phosphorus donor from P_4 to the nickel atom results in the nitrogen donor of the $(Ph_2PCH_2CH_2)_3N$ ligand breaking free from the coordination sphere. This allows the metal atom to obey the 18 electron rule both in the starting material and in the product. Reaction of P_4 with $(Ph_3P)_3RhCl$ gives $(Ph_3P)_2RhCl(\eta^2\text{-}P_4)$ in which two phosphorus atoms of the P_4 ligand interact with the metal atom (Fig. 2.16b). The starting complex contains a 16 electron rhodium(I) centre, but the product obeys the eighteen electron rule.

Not all reactions of white phosphorus with transition metal complexes lead to simple coordination products. The reactions shown in Fig. 2.17 illustrate a range of observed pathways. Disruption of the tetrahedral P_4 cluster occurs, in the simplest case through the substitution of a phosphorus atom for an isolobal transition metal fragment such as $\{Cp^*Cr(CO)_2\}$ or $\{Co(CO)_3\}$. Stepwise substitution may occur as is observed in the reaction of P_4 with $Co_2(CO)_8$. This yields $Co(CO)_3P_3$, $Co_2(CO)_6P_2$, and $Co_3(CO)_9P$ each of which has a tetrahedral cluster core. Reorganization of the phosphorus atoms may occur to generate cyclic ligands of the type $\eta^3\text{-}P_3$, $\eta^4\text{-}P_4$, $\eta^5\text{-}P_5$, and $\eta^6\text{-}P_6$. The reaction of P_4 with $Cp^*Ti(CO)_2$ gives $Cp^*_2Ti_2P_6$ (Fig. 2.17) the structure of which may be described as a *cubane*.

Refer to Section 4.5 for a definition of *isolobality*.

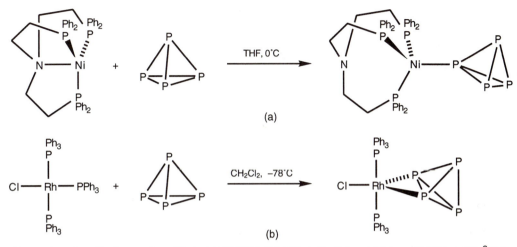

Fig. 2.16 Coordination modes of P$_4$ in (a) Ni{N(CH$_2$CH$_2$PPh$_2$)$_3$}{σ-P$_4$} and (b) *trans*-Rh(PPh$_3$)$_2$Cl{η^2-P$_4$}.

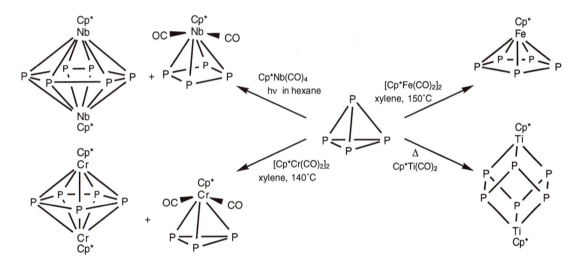

Fig. 2.17 Reactions of white P$_4$ with pentamethylcyclopentadienyl (Cp*) carbonyl derivatives of selected transition metals.

2.4 Arsenic, antimony, and bismuth

In the vapour phase, arsenic exists as tetrahedral As$_4$ molecules. Raman spectroscopic data for the matrix isolated species confirm tetrahedral (T_d) symmetry. At relatively low temperatures, antimony vapour consists mainly of Sb$_4$ molecules. Matrix isolation experiments have confirmed the nature of this species and results indicate the presence of Sb$_2$ and Sb$_3$ in addition to Sb$_4$. Similarly, electronic spectral data for Bi$_n$ in solid neon or argon matrices have shown that Bi$_4$ molecules are present along with Bi$_2$ diatomics. The inaccessibility of As$_4$, Sb$_4$, and Bi$_4$ as room temperature species precludes their use as synthetic precursors. At room temperature and pressure, arsenic, antimony, and bismuth possess structures that are related to the network of puckered six-membered rings observed for black phosphorus.

3 Aspects of the structures of cluster compounds

3.1 Boron

Neutral boranes and hydroborate dianions

exo: This term is applied to an atom or group attached by a localized 2-centre-2-electron bond in a terminal site outside the cluster core.

endo: This term is applied to an atom or molecular fragment which bridges an edge or caps a face of the cluster core; the electrons from the atom or fragment are involved in cluster bonding.

Cluster molecules and ions containing boron constitute the largest group of clusters exhibited by a single *p*-block element. The simplest clusters in terms of an empirical formula are the boranes and borane anions (hydroborates). A borane is a compound which contains only boron and hydrogen; it is also known as a boron hydride. Compounds fall into series according to formula and whether or not the cluster is a single cage or consists of two or more coupled cages. All cages are based upon triangular faced polyhedra; these and some other polyhedra incorporating square faces are drawn on the front inside book-cover.

Classification of borane clusters with representative members of each class

Single cage	*closo*	$[B_nH_n]^{2-}$	$n = 6\text{–}12$
Single cage	*nido*	B_nH_{n+4}	$n = 5, 6, 8, 10, 11$
Single cage	*arachno*	B_nH_{n+6}	$n = 4, 5, 6, 9$
Single cage	*hypho*	B_nH_{n+8}	
Cages coupled by a common B atom		$B_7H_{13} = \{B_5H_8\}\{B_2H_5\}$	
Cages coupled by an *exo*-B–B bond $\{nido\}_2$		$\{B_nH_{n+3}\}_2$	$n = 5, 10$
Cages coupled by an *exo*-B–B bond $\{arachno\}_2$		$\{B_nH_{n+5}\}_2$	$n = 4$
Cages coupled by a shared B–B edge		$B_{12}H_{16}, B_{13}H_{19}, B_{14}H_{18}, B_{14}H_{20}, B_{16}H_{20}, B_{18}H_{22}$	
Cages coupled via B_3-faces		$B_{20}H_{16}$	

Nomenclature for borane clusters

The name should reveal the number of boron atoms, the number of hydrogen atoms, and the charge. The number of boron atoms is given by a Greek prefix except for the use of Latin nona- (nine) and undeca- (eleven). The number of hydrogen atoms appears as an Arabic numeral after the written name. For an ion, the charge is given at the end of the name. Strictly, the class of cluster should also be included.

B_4H_{10}	tetraborane(10)	*arachno*-tetraborane(10)
B_5H_9	pentaborane(9)	*nido*-pentaborane(9)
B_5H_{11}	pentaborane(11)	*arachno*-pentaborane(11)
B_6H_{10}	hexaborane(10)	*nido*-hexaborane(10)
$[B_6H_6]^{2-}$	hexahydrohexaborate(2–)	hexahydro-*closo*-hexaborate(2–)
$B_{10}H_{14}$	decaborane(14)	*nido*-decaborane(14)
$B_{10}H_{16} = \{B_5H_8\}_2$	1,1'- or 1,2'- or 2,2'-decaborane(16)	*conjuncto*-1,1'- or 1,2'- or 2,2'-decaborane(16)

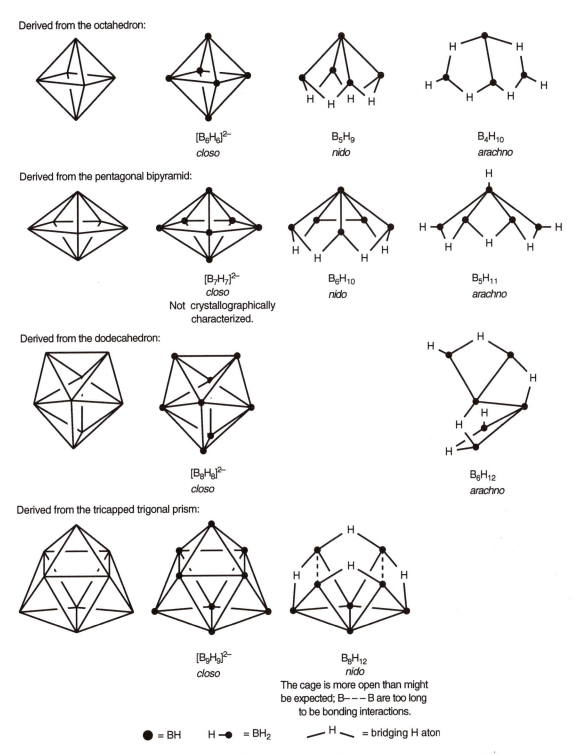

Derived from the octahedron:

[B₆H₆]²⁻
$[B_6H_6]^{2-}$
closo

B₅H₉
nido

B₄H₁₀
arachno

Derived from the pentagonal bipyramid:

$[B_7H_7]^{2-}$
closo
Not crystallographically
characterized.

B_6H_{10}
nido

B_5H_{11}
arachno

Derived from the dodecahedron:

$[B_8H_8]^{2-}$
closo

B_6H_{12}
arachno

Derived from the tricapped trigonal prism:

$[B_9H_9]^{2-}$
closo

B_8H_{12}
nido
The cage is more open than might
be expected; B– – –B are too long
to be bonding interactions.

● = BH   H—● = BH₂   ⌢H⌢ = bridging H atom

Fig. 3.1 Structures of *closo*-hydroborate dianions and *nido*- and *arachno*-boranes possessing single cages.

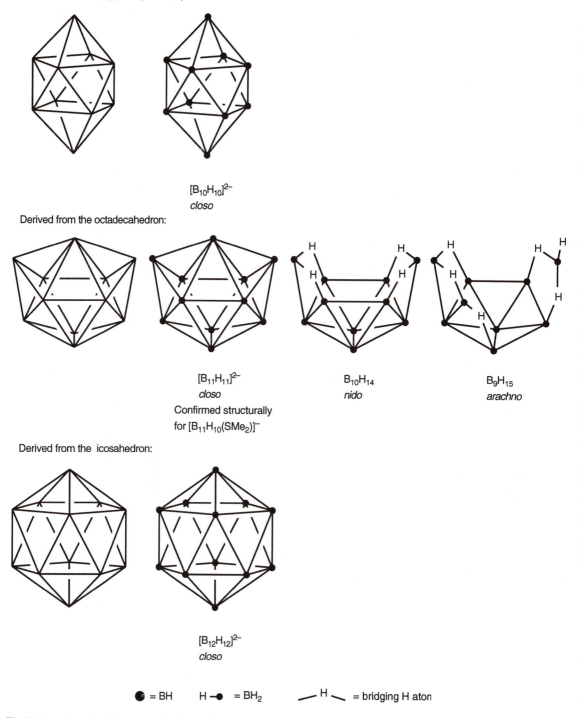

Derived from the bicapped square antiprism:

$[B_{10}H_{10}]^{2-}$
closo

Derived from the octadecahedron:

$[B_{11}H_{11}]^{2-}$
closo
Confirmed structurally
for $[B_{11}H_{10}(SMe_2)]^-$

$B_{10}H_{14}$
nido

B_9H_{15}
arachno

Derived from the icosahedron:

$[B_{12}H_{12}]^{2-}$
closo

● = BH   H—● = BH$_2$ —H— = bridging H atom

Fig. 3.1 (continued)   Structures of *closo*-hydroborate dianions and *nido*- and *arachno*-boranes possessing single cages.

The structures of the *closo*-hydroborate dianions and neutral *nido*- and *arachno*-boranes are given in Fig. 3.1. Each of the dianions $[B_nH_n]^{2-}$ exhibits a closed (*closo*) deltahedral cage. Compounds of type B_nH_{n+4} and B_nH_{n+6} possess open structures, the former being related to the closed deltahedron by the loss of one vertex (*nido*-cluster) and the latter related by the removal of two vertices (*arachno*-cluster). In each of the $[B_nH_n]^{2-}$, B_nH_{n+4}, and derivative clusters, every boron atom carries one terminal hydrogen atom. For a boron atom of low connectivity in a cluster of type B_nH_{n+6}, two terminal hydrogen atoms may be attached. Clusters consisting of two or more linked cages are called *conjuncto*-boranes.

Of the borane clusters consisting of coupled cages, those which are coupled via a localized 2-centre-2-electron *exo*-B–B bond are readily identified in terms of the constituent single clusters. $\{B_5H_8\}_2$ is derived from two B_5H_9 clusters, with a B–B bond replacing one terminal B–H bond on each cluster. Since each B_5H_9 cluster possesses two types of boron atom, apical and basal, there are three possible modes of coupling for $\{B_5H_8\}_2$ (Fig. 3.2). Similar couplings are seen for $\{B_4H_9\}_2$ (derived from two B_4H_{10} molecules), for $\{B_4H_9\}\{B_5H_8\}$ (derived from coupling B_5H_9 with B_4H_{10}), and for $\{B_{10}H_{13}\}_2$ (derived from two $B_{10}H_{14}$ molecules). The triple cluster $\{B_{10}H_{13}\}\{B_{10}H_{12}\}\{B_{10}H_{13}\}$ is related to $\{B_{10}H_{13}\}_2$; 546 possible isomers (including enantiomers) may be drawn for the triple cluster.

The class names *closo*, *nido*, and *arachno* are derived from the Latin *clovis* for cage (Greek $\kappa\lambda\omega\beta o\varsigma$), Latin *nidus* for nest, and Greek $\alpha\rho\alpha\chi\nu\eta$ for web, respectively.

See Section 4.6 for details of the relationships between the classes.

Coupled borane cages are called *conjuncto*-clusters.

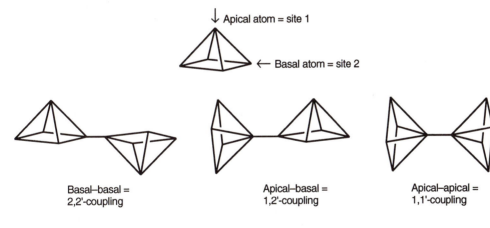

↓ Apical atom = site 1

← Basal atom = site 2

Basal–basal =
2,2'-coupling

Apical–basal =
1,2'-coupling

Apical–apical =
1,1'-coupling

Fig. 3.2 Atom numbering in the core of B_5H_9 and modes of possible coupling in $\{B_5H_8\}_2$. All three isomers are known.

The structure of $\{B_5H_8\}\{B_2H_5\}$ may be described in terms of a B_5H_9 cluster, one BH-unit of which has been replaced by a $\{B_3H_5\}$-unit. This gives rise to the coupling of B_5- and B_3-units by a common boron atom (Fig. 3.3).

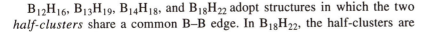

Fig. 3.3 Formal B_n-cage coupling to give the framework of $\{B_5H_8\}\{B_2H_5\}$.

$B_{12}H_{16}$, $B_{13}H_{19}$, $B_{14}H_{18}$, and $B_{18}H_{22}$ adopt structures in which the two *half-clusters* share a common B–B edge. In $B_{18}H_{22}$, the half-clusters are equivalent: two condensed $B_{10}H_{14}$ clusters. There are several ways in which the coupling could occur and in practice two isomers have been characterized. The *anti*-isomer (or *n*-isomer) is centrosymmetric (Fig. 3.4) and the *syn*-isomer (or *iso*-isomer) is non-centrosymmetric.

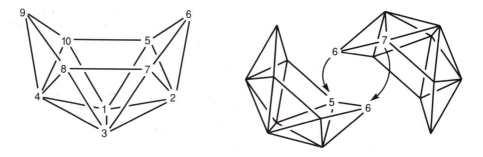

Fig. 3.4   Atom numbering scheme in $B_{10}H_{14}$ and the fusion of two *arachno*-B_{10}-cages to give the centrosymmetric (*anti*) isomer of $B_{18}H_{22}$. In $B_{18}H_{22}$, one B–B edge of each of the original B_{10}-cages is shared.

Carbaboranes

The isoelectronic relationship between a $\{BH\}^-$ and a CH unit makes it possible for the former to be replaced by the latter unit in a borane or hydroborate cluster. Polyhedral clusters which comprise boron and carbon atoms are called *carbaboranes* (Fig. 3.5); the name *carborane* is also commonly used.

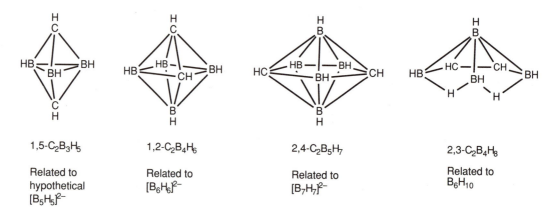

Fig. 3.5   Selected carbaborane clusters. The structurally related and isoelectronic borane or hydroborate dianion is given.

Carbaboranes follow the same classification as boranes and constitute a large group of cluster compounds of which the isomers of *closo*-$C_2B_{10}H_{12}$ (Fig. 3.6) are amongst the most widely studied. For a cluster with a C_xB_y-core, it is usual that $y > x$. The paucity of carbon-rich carbaboranes ($x > y$) may be understood if one remembers that a $\{CH\}$-unit replaces a $\{BH\}^-$- or $\{BH_2\}$-unit. Thus, beginning with a *closo*-hydroborate dianion, it only takes the introduction of two carbon vertices to render the cluster neutral. Cationic clusters are rare in group 13 or 14. Starting from a *nido*-borane B_nH_{n+4}, it is possible to introduce up to four carbon atoms without affecting the overall molecular charge. The structure of one of the two isomers of $C_4Me_4B_4H_4$ is related to B_8H_{12}, and $C_4Me_4B_2H_2$ is related to B_6H_{10}. Even though a range of carbaboranes with a relatively high carbon content may be formulated on the basis of known *nido-* and *arachno-*boranes, in practice the number of known clusters of this type is small.

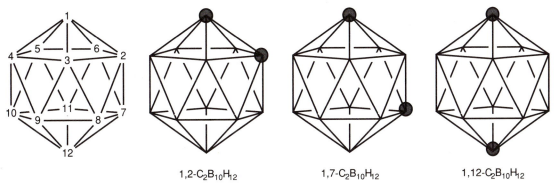

1,2-$C_2B_{10}H_{12}$ 1,7-$C_2B_{10}H_{12}$ 1,12-$C_2B_{10}H_{12}$

Fig. 3.6   Numbering scheme for the icosahedron, and the three isomers of *closo*-$C_2B_{10}H_{12}$.

Isoelectronic species

When two or more molecules, ions, or molecular fragments possess the same number of valence *and* core electrons, they are said to be *isoelectronic*. This term is, however, often used to refer to species which simply possess the same number of valence electrons.

With respect to the number of valence electrons available, the following fragments are isoelectronic with each other:

$$\{BH\}^-, \{CH\}, \{CMe\}, \{NH\}^+, \{AlMe\}^-, \{SiEt\}$$

The following tetrahedral clusters possess the same number of valence electrons within the cluster and are regarded as being isoelectronic with each other:

$$P_4, As_4, Bi_4, Sb_4, Si_4H_4 \text{(hypothetical)}, C_4{}^tBu_4$$

Heteroboranes

The substitution into boranes of cluster units isoelectronic with $\{BH\}^-$ may be extended to fragments other than $\{CH\}$. As will be described in Chapter 4, it is the number of available *valence* electrons possessed by a fragment which controls its ability to mimic a monoborane vertex in a cluster. Thus the CH-for-$\{BH\}^-$ replacement may be extended to CR and SiR (R = alkyl or aryl). Similarly, a Sn or Pb atom that bears an *exo*-lone pair of electrons may replace a neutral $\{BH\}$-unit, and an $\{NH\}$-unit or an S atom may replace a $\{BH\}^{2-}$-unit. This is no different from replacing a $\{CH_2\}$-unit in an alkane by an $\{NH\}$ or $\{O\}$ to generate a secondary amine or ether, respectively. Main-group fragment substitutions of this type lead to a wide range of heteroborane clusters and examples are shown in Fig. 3.7.

The incorporation of an $\{AlMe\}$ fragment in place of a $\{BH\}$ vertex in $[B_{12}H_{12}]^{2-}$ is shown in Fig. 3.7. Likewise, AlB_4H_{11} is structurally related to *arachno-*B_5H_{11}. The incorporation of a $\{GaR\}$ or $\{InR\}$ fragment into a borane or carbaborane cluster gives rise to gallaboranes, gallacarbaboranes, indaboranes, or indacarbaboranes such as 2-GaB_3H_{10} (Fig. 3.8), 1-Me-1-E-2,3-$C_2B_4H_6$ (E = Ga or In, Fig. 3.7) or 1-Et-1-Ga-2,3-$C_2B_9H_{11}$. A naked thallium atom is present in $[3\text{-Tl-}1,2\text{-C}_2\text{B}_9\text{H}_{11}]^-$ although the rather long Tl–B (range 2.66 Å to 2.74 Å) and Tl–C distances (2.91 Å and 2.92 Å) suggest that a better formulation is as the ion-pair $[\text{Tl}]^+ [1,2\text{-C}_2\text{B}_9\text{H}_{11}]^{2-}$.

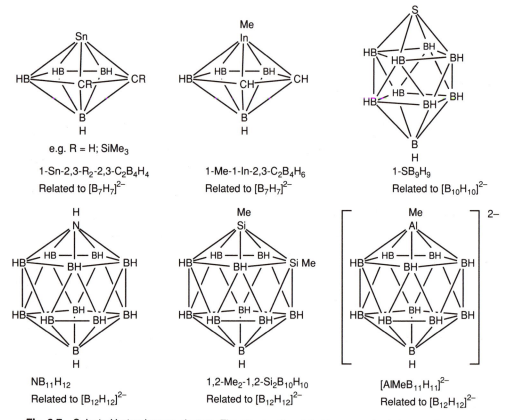

e.g. R = H; SiMe₃

1-Sn-2,3-R₂-2,3-C₂B₄H₄
Related to $[\text{B}_7\text{H}_7]^{2-}$

1-Me-1-In-2,3-C₂B₄H₆
Related to $[\text{B}_7\text{H}_7]^{2-}$

1-SB₉H₉
Related to $[\text{B}_{10}\text{H}_{10}]^{2-}$

NB₁₁H₁₂
Related to $[\text{B}_{12}\text{H}_{12}]^{2-}$

1,2-Me₂-1,2-Si₂B₁₀H₁₀
Related to $[\text{B}_{12}\text{H}_{12}]^{2-}$

$[\text{AlMeB}_{11}\text{H}_{11}]^{2-}$
Related to $[\text{B}_{12}\text{H}_{12}]^{2-}$

Fig. 3.7 Selected heteroborane clusters. The structurally related borane or hydroborate dianion is given.

The descriptor *commo* indicates that the specified atom is sited at the point of fusion of two clusters which share a common atom.

Incorporation of silicon atoms into borane clusters has the scope to provide a range of compounds related to carbaboranes since {SiR} ≡ {CR}. Despite this theoretical prospect, only a few such compounds are known, e.g. 1,2-Me₂-Si₂B₁₀H₁₂ (Fig. 3.7) and *commo*-3,3'-(3-Si-1,2-C₂B₉H₁₁)₂ (Fig. 3.9).

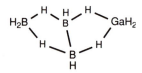

Fig. 3.8 2-GaB₃H₁₀, an analogue of B₄H₁₀.

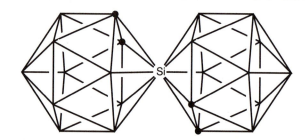

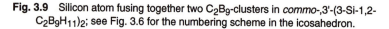

Fig. 3.9 Silicon atom fusing together two C₂B₉-clusters in *commo*-,3'-(3-Si-1,2-C₂B₉H₁₁)₂; see Fig. 3.6 for the numbering scheme in the icosahedron.

Clusters of boron other than the polyhedral boranes

The boron halides B_4Cl_4, B_8Cl_8, and B_9Cl_9 exhibit closed deltahedral B_n cores (Fig. 3.10). B_9Br_9 is isostructural with B_9Cl_9. Other members of this series are B_nCl_n for n = 10–12, B_nBr_n for n = 7–10, B_nI_n for n = 8 and 9, as well as several coupled clusters, e.g. $\{B_9Br_8\}_2$. No fluoride derivatives are known. It is significant that these molecules are *neutral* and yet possess closed cage structures. The dianions $[B_nX_n]^{2-}$ (n = 6, 9; X = Cl, Br, I) and $[B_nX_n]^{2-}$ (n = 10, 12; X = Cl, Br) also exhibit closed deltahedral B_n-cluster cores. This contrasts with the boron hydrides where *closo*-species are observed for dianions $[B_nH_n]^{2-}$ and neutral analogues are not known. On the other hand, the *neutral* $B_4{}^tBu_4$ cluster is stable and a closed tetrahedral cage analogous to that in B_4Cl_4 has been crystallographically confirmed. The mean B–B distances in B_4Cl_4 and $B_4{}^tBu_4$ are equal (1.71 Å).

Classical structures (structures with 2-centre-2-electron bonds) are not usually associated with boranes or carbaboranes but there are occasions when a classical structure is favoured. The tendency for a boron atom to participate in cluster formation arises from a need to use its three valence electrons as effectively as possible. If a π-electron donor is present, the formation of a deltahedral cluster skeleton may not be necessary. This is the case in $C_2H_2B_4(N^iPr_2)_4$ (Fig. 3.11). The lone pair of electrons on each nitrogen atom is donated into the empty $2p$ orbital on the adjacent boron atom.

> Remember that a boron atom has fewer valence electrons than valence orbitals.

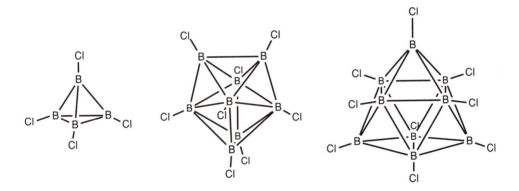

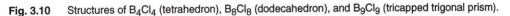

Fig. 3.10   Structures of B_4Cl_4 (tetrahedron), B_8Cl_8 (dodecahedron), and B_9Cl_9 (tricapped trigonal prism).

Although a large group of compounds exhibiting boron–nitrogen rings exists, related cluster systems are not well represented. The same is true for B–P, B–As, B–O, B–S, and B–Se systems. Both the diamond-type (cubic) and hexagonal forms of boron nitride, $(BN)_x$, possess lattice structures. From the former, it is clear that an adamantane-type unit is feasible for a combination of boron and nitrogen atoms but it is not exploited at the molecular level. An adamantane-like structure is observed for the anion $[B_4S_{10}]^{8-}$ (Fig. 3.12) and has been structurally characterized for the lead(II) salt; each boron atom is 4-coordinate.

A cubane-like cluster with two open edges is observed for $[MeNBCl_2]_4$ (Fig. 3.13). Related to this is $[{}^tBuPB(Cl)CH_2B(Cl)P^tBu]_2$ in which the open B----B edges (see Figs. 3.13 and 5.10) of the cubane-like B_4P_4-core are bridged by methylene groups.

> The *adamantane* structure is defined in Fig. 3.19.

> The term *cubane* is used to describe a molecule that exhibits a cubic-core of atoms. The archetypal molecule is C_8H_8.

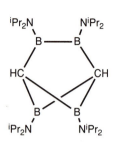

Fig. 3.11 Classical bicyclic structure of $C_2H_2B_4(N^iPr_2)_4$. Compare this to $C_2B_4H_6$ in Fig. 3.5.

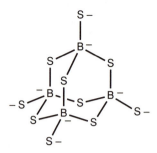
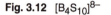

Fig. 3.12 $[B_4S_{10}]^{8-}$

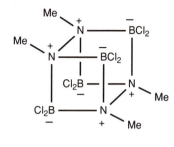

Fig. 3.13 $[MeNBCl_2]_4$.

3.2 Aluminium

Although aluminium is below boron in the periodic table, it does not form many clusters that are analogous to boranes. The dianion $[Al_{12}{}^iBu_{12}]^{2-}$ is a rare example of a cluster in which all the core atoms are aluminium atoms and it illustrates the potential for aluminium to behave like boron in a deltahedral cluster; $[Al_{12}{}^iBu_{12}]^{2-}$ possesses an icosahedral cage like $[B_{12}H_{12}]^{2-}$.

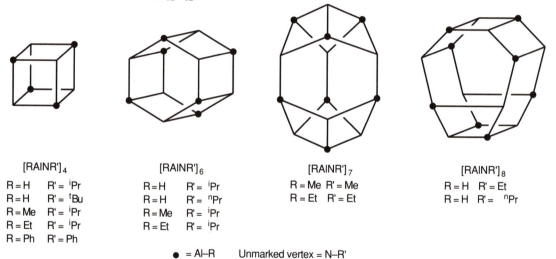

$[RAlNR']_4$	$[RAlNR']_6$	$[RAlNR']_7$	$[RAlNR']_8$
R = H R' = ^iPr	R = H R' = ^iPr	R = Me R' = Me	R = H R' = Et
R = H R' = ^tBu	R = H R' = ^nPr	R = Et R' = Et	R = H R' = ^nPr
R = Me R' = ^iPr	R = Me R' = ^iPr		
R = Et R' = ^iPr	R = Et R' = ^iPr		
R = Ph R' = Ph			

● = Al–R   Unmarked vertex = N–R'

Fig. 3.14 Structures of some iminoalane clusters, $[RAlNR']_n$.

The Lewis acidity of an AlR_3 (R = H, alkyl, or aryl) group allows the formation of adducts with Lewis bases. This general reactivity extends to cluster formation between {RAl} and {NR'} fragments. The degree of oligomerization depends upon the synthetic conditions and on the steric requirements of the *exo*-substituents. Examples of tetrameric, hexameric, heptameric, and octameric iminoalane clusters are shown in Fig. 3.14 and in each case, each aluminium and nitrogen atom is 4-coordinate forming one *exo*-bond and three skeletal interactions. The tetramer forms a cubic cage (cubane) and the hexamer is a hexagonal prism. The heptamer and octamer are not regular polyhedra; like the smaller clusters, the cages of [RAlNR']$_n$ for $n = 7$ or 8 are constructed of fused Al$_2$N$_2$- and Al$_3$N$_3$-rings. The first aluminaphosphacubane [RAlPR']$_4$ (R = ^{i}Bu and R' = SiPh$_3$) has been structurally characterized and is similar to the iminoalane cubane shown in Fig. 3.14.

The inclusion of {ER$_2$}-units (E = Al or N) causes deviation from a closed cage structure; the presence of two terminal substituents restricts the participation of the ER$_2$-unit in cage bonding. The structures of the clusters (HAlN^{i}Pr)$_2$(H$_2$AlNH^{i}Pr)$_3$, (ClAl)$_4$(NMe)$_2$(NMe$_2$)$_4$, and (MeAlNMe)$_6$-(Me$_2$AlNHMe)$_2$ are shown in Fig. 3.15.

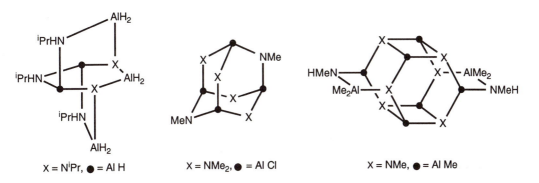

X = N^{i}Pr, ● = Al H   X = NMe$_2$, ● = Al Cl   X = NMe, ● = Al Me

Fig. 3.15   Structures of (HAlN^{i}Pr)$_2$(H$_2$AlNH^{i}Pr)$_3$, the adamantane-like (see Section 3.5) (ClAl)$_4$(NMe)$_2$(NMe$_2$)$_4$, and (MeAlNMe)$_6$(Me$_2$AlNHMe)$_2$. In general, ER$_2$ groups are sited along open edges of the cage; an exception is observed for two NMe groups in (ClAl)$_4$(NMe)$_2$(NMe$_2$)$_4$.

3.3 Gallium and indium

The number of *p*-block clusters involving gallium or indium atoms is limited. (MeGaNMe)$_6$(Me$_2$GaNHMe)$_2$ is isostructural with its aluminium counterpart (Fig. 3.15) but analogues of the iminoalanes shown in Fig. 3.14 are not (as yet) known.

The anions [Ga$_4$S$_{10}$]$^{8-}$, [Ga$_4$Se$_{10}$]$^{8-}$, [In$_4$S$_{10}$]$^{8-}$, and [In$_4$Se$_{10}$]$^{8-}$ are isostructural with [B$_4$S$_{10}$]$^{8-}$ (Fig. 3.12). The same structural unit is observed in [E$_{10}$S$_{16}$(SPh)$_4$]$^{6-}$ for E = Ga or In (Fig. 3.16). In forming these extended adamantane-like structures, the thiolate-sulfides of the group 13 elements parallel the behaviour of the d^{10} metals zinc and cadmium, for example in [Zn$_{10}$S$_4$(SPh)$_{16}$]$^{4-}$ or [Cd$_{17}$S$_4$(SPh)$_{28}$]$^{2-}$. The relationship follows from the relative positions of Zn, Cd, Ga, and In in the periodic table; zinc(II) is isoelectronic with gallium(III), and cadmium(II) is isoelectronic with indium(III).

3.4 Thallium

A usual feature of a thallium atom in a cluster is the lack of a terminal substituent. This may be attributed to the *inert pair effect*. Naked thallium atoms are incorporated into deltahedral Zintl ions such as [TlSn$_8$]$^{3-}$ and [TlSn$_9$]$^{3-}$ (see Section 3.7).

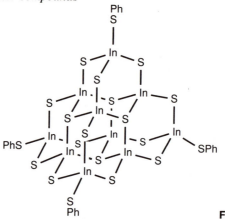

Fig. 3.16 $[In_{10}S_{16}(SPh)_4]^{6-}$.

Inert pair effect

The two *s*-electrons of a heavy *p*-block element may remain as non-valence electrons. This leads to a change in the characteristic oxidation state from *n* to (*n*–2) for the element.

e.g. Tl (group 13) shows a tendency to exhibit oxidation state +1 whereas Al (higher in the same group) exhibits oxidation state +3.

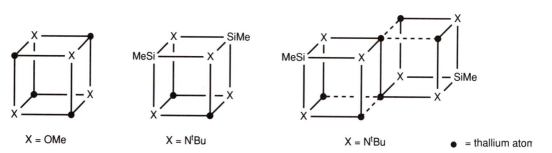

X = OMe   X = N^{t}Bu   X = N^{t}Bu   ● = thallium atom

Fig. 3.17   Structures of $Tl_4(OMe)_4$, $Tl_2(MeSi)_2(N^tBu)_4$, and $Tl_6(MeSi)_2(N^tBu)_6$.

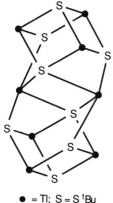

● = Tl; S = S ^{t}Bu

Fig. 3.18 $Tl_8(S^tBu)_8$.

In a series of thallium containing clusters, thallium acts as a Lewis acid. The cube is the structural motif in $Tl_4(OMe)_4$, $Tl_2(MeSi)_2(N^tBu)_4$, and $Tl_6(MeSi)_2(N^tBu)_6$ (Fig. 3.17). In $Tl_4(OMe)_4$ and $Tl_2(MeSi)_2(N^tBu)_4$, each thallium atom is 3-coordinate. In $Tl_6(MeSi)_2(N^tBu)_6$, the double-cubane structure is supported by a short Tl–Tl edge of length 3.16 Å. This is considered to be a bonding interaction; the average *intra*-cube Tl---Tl distance is 3.54 Å. Cubane-residues are also apparent in $Tl_8(S^tBu)_8$ (Fig. 3.18).

3.5 Carbon

The number of organic molecules with fused-cyclic carbon skeletons is vast and will not be explored here save for a few relatively small molecules, the structures of which are related to those observed for cluster molecules of other *p*-block elements.

Adamantane, $C_{10}H_{16}$, (Fig. 3.19) is a tricyclic molecule and the structure resembles a portion of the diamond lattice (Fig. 2.4). There are two carbon sites: in one site the carbon atom is 3-coordinate within the cage and has one terminal substituent and in the second site the carbon atom is 2-coordinate within the cage and has two terminal substituents. Each carbon atom is in a tetrahedral environment. Adamantane-like clusters occur regularly within the *p*-block and the differentiation between the two cage sites not only permits the incorporation of {ER}- and {ER$_2$}-units but also allows the involvement of atoms with different numbers of available valence electrons (Fig. 3.19). More complex cages with recognizable adamantane units are diadamantane (congressane), triadamantane, and tetraadamantane (Fig. 3.20). Collectively these are examples of *diamondoid* hydrocarbons.

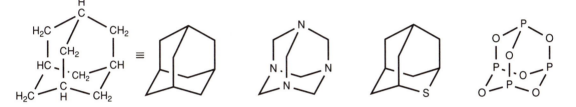

Fig. 3.19 Structure of adamantane, $C_{10}H_{16}$, and three compounds with the adamantane cage: hexamethylene triamine (1,3,5,7-tetrazatricyclo[3.3.1.1^{3,7}]decane), thiaadamantane, and phosphorus(III) oxide, P_4O_6. Note that group 15 atoms replace (are isoelectronic with) {CH} units and group 16 atoms replace (are isoelectronic with) {CH$_2$}-units.

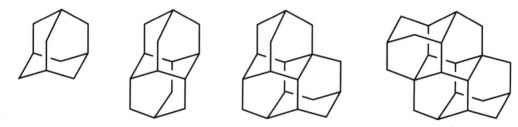

Fig. 3.20 Structures of the diamondoid hydrocarbons adamantane, diadamantane, triadamantane, and tetraadamantane. The *anti*-isomer of tetraadamantane is shown.

In the diamondoid cages, each carbon atom is 4-coordinate and tetrahedrally sited. In cubane, C_8H_8, (Fig. 3.21) and its derivatives, each 4-coordinate atom appears to be in a strained environment since the *intra*-cube C–C–C bond angles are fixed at 90°. The carbon atom can tolerate this structural constraint and, moreover, can accommodate further strain as evidenced in small ring systems such as cyclopropane and in a tetrahedrane cluster, both of which have C–C–C bond angles of 60°. The tetrahedral cluster requires bulky substituents for stability and crystallographic confirmation of the structure comes from a study of $C_4{}^tBu_4$. The C–C bond distances of 1.48 Å compare with a typical C–C single bond length of 1.54 Å. Note that, although not isoelectronic with one another, $C_4{}^tBu_4$ and $B_4{}^tBu_4$ are isostructural. The B_4-core is significantly larger than the C_4-core with a mean B–B distance of 1.71 Å.

Fig. 3.21 Cubane

Benzvalene   Prismane
Fig. 3.22 Isomers of C_6H_6 with cage structures.

Isomers of C_6H_6 include hexadiynes, hexatetraene, *tris*(methylene)cyclo-propane, *bis*(methylene)cyclobutane, fulvene and Dewar benzene:

fulvene   Dewar benzene

Only prismane and benzvalene have structures which might be considered to be clusters of carbon atoms.

Benzene is a molecule familiar to all chemists but its isomers benzvalene and prismane (Fig. 3.22) are perhaps less well known. Whereas an inorganic chemist might see these molecules as possessing cluster structures, an organic chemist would no doubt consider the so-called cluster to be a series of fused carbon rings and similarly for all the carbon-based molecules mentioned in this section. The systematic name for benzvalene bears this out: tricyclo[3.1.0.0²,⁶]hex-3-ene. Structural analyses of the prismane C_6Me_6 in the gas phase and derivatives of prismane such as 1-MeCO₂-2,3,5,6-Me₄-4-Ph-C_6 in the solid state have confirmed the trigonal prismatic C_6-cluster core.

3.6   Silicon

Most of the cyclic silanes $(SiR_2)_n$ are mono- or bicyclic molecules. An adamantane-like structure is observed for $Si_{10}Me_{16}$ and silanes with related structures are $Si_{11}Me_{18}$ and $Si_{13}Me_{22}$ (Fig. 3.23). $P_4(SiMe_2)_6$ is related to $Si_{10}Me_{16}$; the {SiMe}-units of $Si_{10}Me_{16}$ are replaced by phosphorus atoms. Replacing the {SiMe₂}-units by {PMe}-groups gives $(SiMe)_4(PMe)_6$. Oxygen atoms may occupy the edge sites of the cage, e.g. in $(Si^tBu)_4O_6$. A silicon analogue of a substituted cubane, $Si_8(Si^tBuMe_2)_8$, (Fig. 3.24) has been characterized.

● = SiMe₂
Unmarked vertex = SiMe

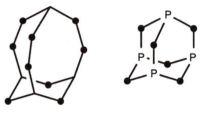

Fig. 3.23   Structures of $Si_{11}Me_{18}$, $Si_{13}Me_{22}$, and $P_4(SiMe_2)_6$.

Silicon–oxygen single bonds are strong and their favourability is evident when one considers the great abundance of silicate minerals and the stability of silicone polymers. A series of discrete clusters containing a Si_8O_{12}-core (Fig. 3.24) is known. Interest in these siloxane cages arises from their potential use as precursors for synthetic silica-based materials comprising linked Si_8O_8-clusters. The isoelectronic principle indicates that analogous silazanes should be accessible by replacing O atoms with {NR}-units. As with the siloxanes, simple ring compounds outnumber cluster structures in silazane chemistry; an example of a cage silazane is $(Si^nC_8H_{17})_8(NH)_{12}$ with a structure resembling that shown for the oxide in Fig. 3.24.

Fig. 3.24 Si_8-based clusters.

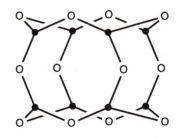

● = Si(Si^tBuMe_2)   ● = SiH, [SiO]⁻, Si(OMe), or SiCl

Various theoretical studies have addressed the question of the relative stabilities of silicon analogues of isomers of C_6H_6. The fact that a planar benzene-like structure for Si_6H_6 would not be preferable is consistent with the fact that Si=Si double bonds are not favoured. There is some indication that a prismane-like structure might be adopted preferentially. However, the exercise remains a hypothetical one since Si_6H_6 has not been isolated. Similarly, theorists suggest that a tetrahedrane-like Si_4H_4 would be unstable although with a suitable substituent, R, tetrahedral Si_4R_4 might be a realistic goal. Significantly, tetrahedral $[Si_4]^{4-}$ anions (isoelectronic with P_4) are present in alkali metal silicides. In $Cs_4[Si_4]$, the solid state lattice is considered to contain isolated $[Si_4]^{4-}$ anions but only to a limited extent; there is considerable cation–anion interaction. In $K_3Li[Si_4]$, lithium ions link the $[Si_4]^{4-}$ tetrahedra together to form an infinite chain. In $K_7Li[Si_4]_2$, pairs of tetrahedra associate with a lithium ion (Fig. 3.25) and further interactions with the K^+ ions occur in the solid state.

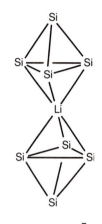

Fig. 3.25 $[(Si_4)_2Li]^{7-}$ unit in $K_7Li[Si_4]_2$.

3.7 Germanium, tin, and lead

Tetrahedral $[E_4]^{4-}$ clusters are found for E = Ge, Sn, and Pb, for example in sodium germide. Isostructural with these cages are the heteroatomic dianions $[Bi_2Sn_2]^{2-}$ and $[Sb_2Pb_2]^{2-}$; the replacement of two group 14 by group 15 elements reduces the overall charge by two, thereby maintaining isoelectronic species. A series of group 14 homoatomic anions (known as *Zintl ions* after Eduard Zintl) is shown in Fig. 3.26. For a given ion $[E_n]^{x-}$, the cluster will be larger as the group is descended and the atomic radius increases; in $[Sn_5]^{2-}$ the equatorial Sn–Sn distance is 3.10 Å while in $[Pb_5]^{2-}$ the corresponding distance is 3.24 Å. A series of anions $[E_{9-n}E'_n]^{4-}$ for E = Sn, E' = Ge or Pb, and $n = 0$ to 9 is known. $[TlSn_8]^{3-}$ is formally derived from the (unknown) $[Sn_9]^{2-}$ cluster by replacing a tin atom by an isoelectronic thallium(1–) centre. Structurally, $[TlSn_8]^{3-}$ is related to $[Ge_9]^{2-}$. A bicapped square antiprismatic cluster is observed for $[TlSn_9]^{3-}$ for which no isoelectronic group 14 counterpart, $[E_{10}]^{2-}$, is known. In both cluster frameworks there are two different sites available for the thallium atom, one 4-coordinate and one 5-coordinate. In each Zintl ion the thallium atom occupies a capping (4-coordinate) site (Fig. 3.27).

A *Zintl phase* is formed between a very electropositive metal and a less electropositive metal (e.g. between a group 1 metal such as Na and a heavy *p*-block element such as Tl). Discrete *Zintl ions* may be obtained by extraction from Zintl phases (see Section 5.10).

$[Ge_4]^{4-}$
$[Sn_4]^{4-}$
$[Pb_4]^{4-}$

$[Sn_5]^{2-}$
$[Pb_5]^{2-}$

$[Ge_9]^{2-}$

$[Ge_9]^{4-}$
$[Sn_9]^{4-}$
$[Pb_9]^{4-}$

Fig. 3.26 The structures of the group 14 anionic clusters $[E_4]^{2-}$ (tetrahedron), $[E_5]^{2-}$ (trigonal bipyramid), $[E_9]^{2-}$ (tricapped trigonal prism), and $[E_9]^{4-}$ (monocapped square antiprism).

Fig. 3.27 Structures of $[TlSn_8]^{3-}$ and $[TlSn_9]^{3-}$.

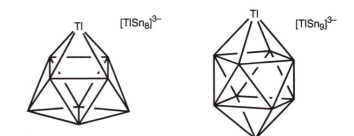

$[TlSn_8]^{3-}$ $[TlSn_9]^{3-}$

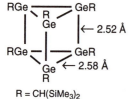

R = CH(SiMe₃)₂

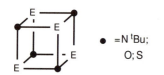

● = GeBr^{t}Bu;
unmarked vertex = Ge^{t}Bu

Fig. 3.28 Structures of $Ge_6\{CH(SiMe_3)_2\}_6$ and $Ge_8{}^tBu_8Br_2$.

Until 1989, polycyclic germanes had not been reported. The compound $Ge_6\{CH(SiMe_3)_2\}_6$ (Fig. 3.28) is related to prismane. The choice of the *exo*-substituent is critical to the stability of the polyhedral cluster. Although octagermanacubane clusters are not known, the tetracyclic molecule $Ge_8{}^tBu_8Br_2$ is. Its structure (Fig. 3.28) consists of a network of fused 4- and 5-membered rings.

As for the earlier members of group 14, germanium atoms are involved in a variety of compounds related to adamantane. The structure of $[Ge_4X_{10}]^{4-}$ (X = S or Se) mimics that of the isoelectronic anion $[E_4S_{10}]^{8-}$ (E = B, Ga, or In). Fig. 3.29 illustrates several clusters which contain germanium, the structures of which are clearly related to those of adamantane. As shown in Fig. 3.19, the preference of an atom or group for a particular site in the framework of the cluster should be considered in terms of the isoelectronic principle. The adamantane framework is also established in the chemistry of tin; the structures of $P_4(SnEt_2)_6$ and $(SnMe)_4E_6$ (E = S or Se) are related to those of $P_4(SiMe_2)_6$ (Fig. 3.23) and $(GeCF_3)_4S_6$ respectively.

Fig. 3.29 Germanium-containing compounds which adopt an adamantane-type of cage.

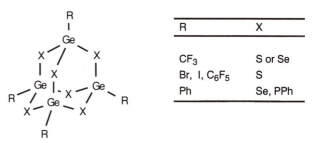

R	X
CF₃	S or Se
Br, I, C₆F₅	S
Ph	Se, PPh

The cubane structure (Fig. 3.30) is adopted by several germanium, tin, or lead containing clusters, the polycyclic cage being stabilized by Lewis acid–Lewis base interactions. The compounds $[EN^tBu]_4$ (E = Ge, Sn, or Pb), $Ge_3Sn(N^tBu)_4$, $Sn_2Pb_2(N^tBu)_4$, $Sn_4O(N^tBu)_3$, and $Sn_4S(N^tBu)_3$ are all representative of this family of clusters.

Iminostannylene clusters, $[SnNR]_4$, with a range of substituents, R, have been prepared but if the steric requirements of R are small, the degree of oligomerization increases and polymeric products $[SnNR]_\infty$ result. Cubic and hexagonal prismatic cores are observed for the tin alkoxides shown in Fig. 3.31. Each cluster comprises either an $\{Sn_4O_4\}$- or $\{Sn_6O_6\}$-core and is stabilized by bidentate *O,O'*-donor ligands. Each bidentate ligand bridges diagonally across a Sn_2O_2-square face. In the cubane, two square-faces are open and four are bridged; in the hexagonal prism, all six square-faces are

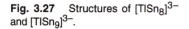

● = N ^{t}Bu;
O; S

Fig. 3.30 Cubanes containing atoms from group 14, E = Ge, Sn, or Pb.

bridged. The *O,O'*-ligands form a zig-zag pattern around the outside of the central Sn_nO_n-cluster core. ^{119}Sn NMR spectroscopy is a useful structural probe in systems of this type. Related to $[^nBuSn(O)(\mu-O_2P(C_6H_{11})_2]_4$ is the double-cubane cluster $[\{^nBuSn(S)(\mu-O_2PPh_2)\}_3O]_2Sn$. The linked cubic-cores are composed of Sn–O and Sn–S interactions (Fig. 3.31).

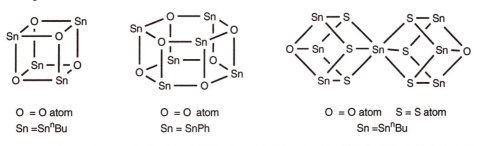

O = O atom   O = O atom   O = O atom   S = S atom
Sn =Sn^{n}Bu   Sn = SnPh   Sn =Sn^{n}Bu

Fig. 3.31 Cluster frameworks (omitting bridging ligands) of the alkoxides $[^nBuSn(O)(\mu-O_2P(C_6H_{11})_2]_4$, $[PhSn(O)(\mu-O_2CC_6H_{11})]_6$, and the related double cubic cluster $[\{^nBuSn(S)(\mu-O_2PPh_2)\}_3O]_2Sn$.

3.8 Phosphorus

In terms of valence electrons, a group 15 atom is isoelectronic with a {CR}-unit. Thus, white phosphorus, P_4, is related to the tetrahedrane $C_4{}^tBu_4$. Several phosphorus-containing clusters are known for which a parent polycyclic-hydrocarbon molecule is easily recognized (Fig. 3.32).

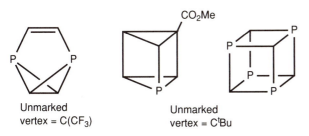

Unmarked
vertex = C(CF$_3$)

Unmarked
vertex = C^{t}Bu

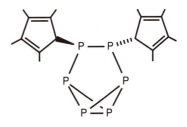

Fig. 3.32 Phosphorus–carbon clusters related to benzvalene, prismane, and cubane by the replacement of {CR}-fragments by P atoms.

Fig. 3.33 $P_6(C_5Me_5)_2$.

Note that, in contrast to some other phosphorus-containing cubanes (see below), the formation of the tetraphosphacubane $[^tBuCP]_4$ (Fig. 3.32) does *not* require the phosphorus atoms to function as Lewis bases and therefore each P atom carries an *exo*-lone pair of electrons. The P–C bond distances of 1.88 Å in $[^tBuCP]_4$ are consistent with single bond character.

The polyphosphane $P_6(C_5Me_5)_2$ (Fig. 3.33) exhibits the same tricyclic framework as benzvalene, although the presence of the {PR}-groups indicates that $P_6(C_5Me_5)_2$ is actually an analogue of the saturated molecule 3,4-dihydro-benzvalene. The presence of a stereochemically active lone pair of electrons on each phosphorus atom leads to the C_5Me_5 substituents adopting a *trans*-configuration.

In Chapter 2, the chemistry of white phosphorus, P_4, was described. The cluster structures of the homoatomic anions $[P_{16}]^{2-}$, $[P_{21}]^{3-}$, and $[P_{26}]^{4-}$ are illustrated in Figs. 2.16 and 2.17. One characteristic structural motif of these

Fig. 3.34 $[P_7]^{3-}$.

anions is that exhibited by $[P_7]^{3-}$ (Fig. 3.34), the simplest cage of this type. The second structural building block is derived from monoclinic phosphorus (Fig. 2.17). At the molecular level, polycyclic phosphorus hydrides and their derivatives exhibit structures which clearly resemble fragments of the lattice of monoclinic phosphorus. Examples are shown in Fig. 3.35 and should be compared with the diagram in Fig. 2.17c.

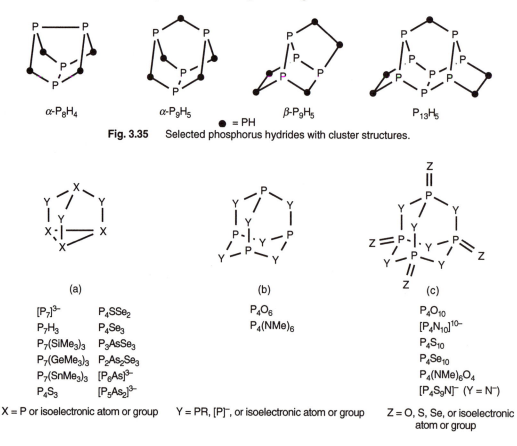

α-P_8H_4 α-P_9H_5 β-P_9H_5 $P_{13}H_5$

● = PH

Fig. 3.35 Selected phosphorus hydrides with cluster structures.

(a)   (b)   (c)

$[P_7]^{3-}$	P_4SSe_2	P_4O_6
P_7H_3	P_4Se_3	$P_4(NMe)_6$
$P_7(SiMe_3)_3$	P_3AsSe_3	
$P_7(GeMe_3)_3$	$P_2As_2Se_3$	
$P_7(SnMe_3)_3$	$[P_6As]^{3-}$	
P_4S_3	$[P_5As_2]^{3-}$	

P_4O_{10}	
$[P_4N_{10}]^{10-}$	
P_4S_{10}	
P_4Se_{10}	
$P_4(NMe)_6O_4$	
$[P_4S_9N]^-$	(Y = N⁻)

X = P or isoelectronic atom or group   Y = PR, [P]⁻, or isoelectronic atom or group   Z = O, S, Se, or isoelectronic atom or group

Fig. 3.36 Important classes of phosphorus compounds which are structurally related to the adamantane-type of cage.

Formal derivation of one class of compound from another in terms of structure does *not* necessarily imply the presence of a related synthetic pathway.

A large number of phosphorus-containing clusters may be formally derived from the $[P_7]^{3-}$ ion. Three basic cluster shapes or derivatives thereof occur amongst such species (Fig. 3.36) and the relationship of each to the adamantane-cage is apparent. Interestingly, not all species that might appear to be possible synthetic targets are known. For instance, group (b) in Fig. 3.36 is rather poorly represented. No doubt new examples will be added to the groups; in group (c), the structure of $[P_4N_{10}]^{10-}$ was elucidated in 1991.

There are many compounds the structures of which lie in between those given in Fig. 3.36. Examples include P_4S_4 and P_4S_5, P_4S_7, P_4S_9, and $P_4O_3S_6$ (Fig. 3.37); a third isomer of P_4S_4 has one *exo*-cyclic sulfur atom. In 1991, P_4S_6 was isolated and structurally characterized; it is *not* isostructural with P_4O_6 but is derived instead from β-P_4S_5. Phosphorus selenides include P_4Se_5

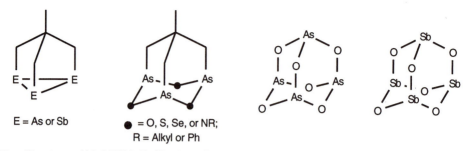

Fig. 3.39 Structures of $MeC(CH_2)_3E_3$ (E = As or Sb), $MeC(CH_2)_3As_3O_3$, $MeC(CH_2)_3As_3(NR)_3$, As_4O_6, and Sb_4O_6.

3.10 Bismuth

Apart from the tetrahedral Zintl ion $[Bi_2Sn_2]^{2-}$ (see Section 3.7), cluster species of bismuth are cationic. The cluster structures of the cations $[Bi_5]^{3+}$, $[Bi_8]^{2+}$, and $[Bi_9]^{5+}$ (Fig. 3.40) are the trigonal bipyramid, square antiprism, and tricapped trigonal prism, respectively. The average Bi–Bi distance in $[Bi_8]^{2+}$ is 3.10 Å which compares well with a value of 3.09 Å in bismuth metal. Similar distances are observed in $[Bi_5]^{3+}$: 3.01 Å for Bi(apex)– Bi(equator) and 3.32 Å for Bi(equator)-Bi(equator). The $[Bi_9]^{5+}$ ion, present in $Bi_{12}Cl_{14}$, exhibits Bi–Bi distances between 3.09 Å and 3.74 Å.

Fig. 3.40 Cationic homoatomic clusters of bismuth.

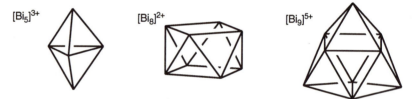

The cluster cation $[Bi_6(OH)_{12}]^{6+}$ is the dominant species in highly acidic aqueous solutions of bismuth(III). The cation possesses an octahedral Bi_6- core, each edge of which is bridged by an hydroxyl group. The internuclear separation of adjacent bismuth atoms is 3.70 Å, a value which falls at the upper end of the distances observed in the cations described above.

3.11 Sulfur, selenium, and tellurium

Unless combined with another element, sulfur atoms do not aggregate to form clusters; likewise for selenium and tellurium. This follows from a consideration of the number of electrons in the valence shell of a group 16 atom (see Section 2.1). The role of sulfur and selenium atoms in supporting cage structures of elements from groups 13 and 14 is illustrated in anions such as $[E_4S_{10}]^{8-}$, (E = B, Ga, or In), $[E_4Se_{10}]^{8-}$ (E = Ga or In), $[E_{10}S_{16}(SPh)_4]^{6-}$ (E = Ga or In), and neutral compounds such as $Tl_8(S^tBu)_8$ (Fig. 3.18), $(GeCF_3)_4Se_6$, (Fig. 3.29), and $[\{^nBuSn(S)(\mu\text{-}O_2PPh_2)\}_3O]_2Sn$ (Fig. 3.31).

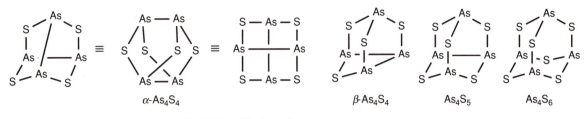

Fig. 3.41   Cluster sulfides of arsenic.

Cluster structures derived from the adamantane framework are exhibited by the arsenic sulfides depicted in Fig. 3.41. The two isomers of As_4S_4 are structurally analogous to α- and β-P_4S_4 (Fig. 3.37). The second and third views of α-As_4S_4 emphasize the fact that (i) the sulfur atoms lie in a plane and (ii) there are two orthogonal As–As bonding interactions (As–As = 2.49 Å). As_4Se_4 is isostructural with As_4S_4 with As–As distances of 2.56 Å. These structures contrast with those adopted by S_4N_4 (Fig. 3.42) and Se_4N_4. In terms of valence electrons, S_4N_4, Se_4N_4, As_4S_4, and As_4Se_4 are equivalent. However, in S_4N_4 and Se_4N_4, it is the nitrogen and not the group 16 atoms that is coplanar. In addition each cage is supported by weak S--S or Se--Se interactions. Each S–N bond in S_4N_4 is of length 1.62 Å and the S--S separation is 2.58 Å; compare this with a value of 2.08 Å for the sum of the covalent radii of two sulfur atoms. In Se_4N_4, the Se-N bond distance is 1.80 Å and the Se--Se separation is 2.76 Å; the sum of the covalent radii of two selenium atoms is 2.34 Å. The higher sulfides of arsenic As_4S_5 and As_4S_6 (Fig. 3.41) are isostructural with P_4S_5 and P_4O_6, respectively. Note that As_4S_6 is *not* isostructural with P_4S_6.

Bond formation amongst atoms of the group 16 elements is common and leads to a large number of cyclic sulfur, selenium, and tellurium molecules and cations. Detailed coverage of these species is beyond the scope of this book. The cations $[Te_6]^{4+}$, $[Te_3Se_3]^{2+}$, and $[Te_2Se_4]^{2+}$ (Figs. 3.43 and 3.44) are often classed as clusters even though the dications are bicyclic species. Each structure is derived from a trigonal prism. For $[Te_6]^{4+}$, an elongated prism is observed. In the two dications, two of the nine edges of the parent trigonal prism are broken (see Section 4.4).

The folded monocyclic cations $[S_8]^{2+}$, $[Se_8]^{2+}$, and $[Te_8]^{2+}$ should be included for completeness. The transannular interaction in $[Te_8]^{2+}$ is 2.99 Å which is *shorter* than the long edges in $[Te_6]^{4+}$; this emphasizes the problems encountered when delineating bonding interactions in such clusters. The simple planar cyclic cations $[E_4]^{2+}$ (E = S, Se, or Te) are not considered here to be clusters.

The *adamantane* structure is defined in Fig. 3.19.

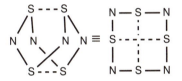

Fig. 3.42   Structure of S_4N_4.

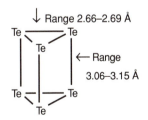

Fig. 3.43   The elongated nature of the prism in $[Te_6]^{4+}$.

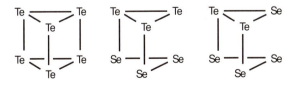

Fig. 3.44   Structures of the cations $[Te_6]^{4+}$, $[Te_3Se_3]^{2+}$, and $[Te_2Se_4]^{2+}$.

4 Bonding in clusters

4.1 Introduction: localized and delocalized bonding

In generating a bonding picture for a simple molecule, the ground state electronic configuration of the central atom is the usual starting point. The number of bonds made by an atom is controlled by the number of valence electrons present and the number of atomic orbitals available for occupancy. For first row elements, the valence atomic orbitals are the $2s$ and the $2p$. Consider a carbon atom. The ground state configuration is $2s^2 2p^2$; after the promotion of one s-electron to a p-orbital to give an excited state (Fig. 4.1) the creation of four bonds follows easily. In general, one of three geometries is expected for a carbon atom: tetrahedral, trigonal planar, or linear. The bonding is easily described in terms of hybridization (Fig. 4.1). Use of sp^3 hybridization allows the formation of four σ-bonds. In an sp^2 or sp hybridized carbon atom, one or two pure p-orbitals, respectively, are reserved for π-bonding. Each bond here is a localized 2-centre-2-electron one.

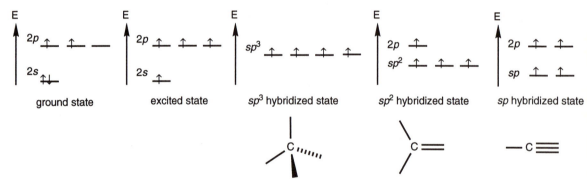

Fig. 4.1 Ground and valence electronic states of the carbon atom in tetrahedral, trigonal planar, and linear geometries.

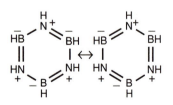

Fig. 4.2 Resonance structures for [HBNH]$_3$.

Delocalization of π-electrons occurs when a π-system is extended over more than one pair of adjacent atoms. Thus, rather than having alternating single and double bonds, the ring structure of benzene exhibits six equivalent C–C bonds due to $p\pi$–$p\pi$ interactions. Borazine, [HBNH]$_3$, (Fig. 4.2) is an inorganic ring which is isoelectronic with benzene. It too is stabilized by π-delocalization. Each boron and nitrogen atom is considered to be sp^2 hybridized; the unhybridized $2p$ atomic orbital (AO) on each nitrogen atom contains a lone pair of electrons whilst the corresponding AO on each boron atom is empty. Thus π-donation from nitrogen to boron occurs and the net result is a six π-electron delocalized system. Unlike the C–C bonds in benzene, the B–N bonds are polarized leaving each nitrogen or boron atom with a net negative or positive charge, respectively. Note that the formal charges drawn in a resonance valence structure are not the same as the net atomic charges in the molecule itself; the Pauling electronegativities of nitrogen and boron are 3.04 and 2.04 respectively. Thus, despite the stabilizing influence of the π-system, $[HBNH]_3$ is susceptible to both nucleophilic (at a boron atom) and electrophilic (at a nitrogen atom) attack.

In borazine, each boron atom acts as a Lewis acid accepting electron density from a neighbouring Lewis base (the nitrogen atom). This tendency is exhibited in many molecules containing boron or other group 13 elements. For example, simple boron trihalides are stabilized by $p\pi-p\pi$ bonding as shown in Fig 4.3. Provided that the $p\pi-p\pi$ bonding is efficient, the Lewis acidity of the boron atom is thereby reduced and attack by external Lewis bases will be hindered. For halogen atoms X in BX_3, halogen to boron $p\pi-p\pi$ bonding follows the order F > Cl > Br > I. In $AlCl_3$, the Lewis acidity of the aluminium atom manifests itself in the formation of the dimer Al_2Cl_6 (Eqn 4.1). Dimerization is accompanied by a change in geometry at the aluminium atoms from trigonal planar to tetrahedral. This change is consistent with going from an aluminium atom which is bonded to three chlorine atoms and has one empty AO (which can only be stereochemically inactive) to one which is connected to four chlorine atoms.

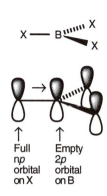

Fig. 4.3 π-Stabilization in BX_3 (X = halogen); π-interactions occur along all three B–X vectors.

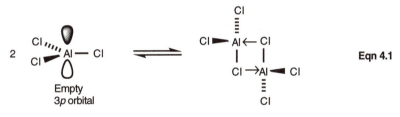

Eqn 4.1

4.2 Small clusters containing carbon. Is localized bonding sufficient?

The bonding in organic cage molecules are usually described simply in terms of localized 2-centre-2-electron interactions. This is a suitable scheme for adamantane and other diamondoid molecules (Figs. 3.19 and 3.20) in which each carbon atom is in a tetrahedral or approximately tetrahedral environment.

Whereas benzene is planar, its isomer benzvalene has a 3-dimensional C_6-cage structure. The valence requirements of each carbon atom in the C_6-framework are satisfied according to the localized bonding scheme shown in Fig. 4.4. Two carbon atoms are formally sp^2 hybridized whilst four are sp^3 hybridized.

In the C_8-cage of cubane, C_8H_8 (Fig. 4.4), each carbon atom occupies a vertex position of the cage and forms one exocyclic (outside the ring or cage) and three endocyclic (part of the ring or cage framework) bonds. For four single bonds, sp^3 hybridization appears to be a satisfactory bonding description for each carbon atom. However, the endocyclic C–C–C bond angles are constrained to be 90°, a significant distortion away from the 109.5° angle associated with sp^3 hybridization. One bonding description that gets around this problem is the concept of *bent bonds*. Using cubane as the example, there are two fixed points of reference. The first is the cubic

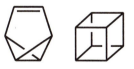

Fig. 4.4 Benzvalene and cubane; each vertex = CH.

A *bent bond* is formed when the centre of bonding electron density between two nuclei lies off the internuclear vector:

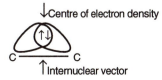

C_8-cage in which endocyclic C–C–C angles are 90°. The second is the assumption that each 4-coordinate carbon atom is sp^3 hybridized. One hybrid orbital is used to bond the exocyclic substituent to the cage carbon atom via a directionalized 2-centre-2-electron bond (Fig. 4.5). This leaves the remaining three sp^3 hybrid orbital lobes pointing *outside* the surface of the cage rather than perfectly along its edges. Interactions between orbitals of adjacent carbon atoms in the cage lead to bent C–C bonds. This concept is used to describe the bonding in small rings such as cyclopropane—it is not unique to clusters.

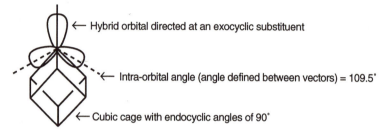

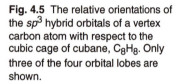

Fig. 4.5 The relative orientations of the sp^3 hybrid orbitals of a vertex carbon atom with respect to the cubic cage of cubane, C_8H_8. Only three of the four orbital lobes are shown.

← Hybrid orbital directed at an exocyclic substituent

← Intra-orbital angle (angle defined between vectors) = 109.5°

← Cubic cage with endocyclic angles of 90°

In the tetrahedrane, C_4R_4 (Fig. 4.6, R = ^{t}Bu) each carbon atom may be considered to be sp^3 hybridized with one hybrid orbital directed towards the exocyclic substituent and each of the three remaining hybrids pointing towards another cage carbon atom but directed outside the C_4-tetrahedron. This leads to six 2-centre-2-electron bent bonds. To a first approximation, this is a useful description. However, the constraints set by the endocyclic C–C–C angles of 60° in tetrahedrane are significantly greater than in cubane. Further, it has been observed that the cage C–C bond lengths in $C_4^tBu_4$ are 1.49 Å compared to a typical value of 1.54 Å for a C–C single bond. ^{13}C NMR spectroscopic chemical shift values are sensitive to changes in the hybridization of a particular carbon atom and shift data for $C_4^tBu_4$ suggest that each cage carbon atom should be considered to be sp hybridized. Assuming that each cluster carbon atom is sp hybridized, a bonding scheme for a tetrahedrane cluster, modelled here by C_4H_4, can be developed as follows, using a hybrid molecular orbital approach. One sp hybrid orbital of each carbon atom of the C_4-cluster core interacts with the $1s$ orbital of a terminal hydrogen atom. This leaves one sp hybrid and two $2p$ orbitals per cluster carbon atom (Fig. 4.7). These three orbitals are termed the *frontier MOs* of the {CH}-group and are available for bonding this unit to the three other {CH}-units to generate the tetrahedral C_4-core of the C_4H_4 molecule. The combination of these orbitals to give six C_4-core bonding MOs is illustrated in Fig. 4.8. The nature of the resultant MOs reflects the fact that of the three frontier orbitals of each cluster unit, the sp orbital points into the centre of the tetrahedral C_4-core and the two p orbitals are tangential to the surface of the core. For the C_4-core of the C_4H_4 molecule, there is one unique MO, Ψ_5, formed by the overlap of the four sp hybrids. There is a triply degenerate set of MOs, Ψ_6 to Ψ_8, one of which is drawn in Fig. 4.8. The highest lying bonding MOs Ψ_9 and Ψ_{10} are degenerate. For the tetrahedrane molecule to be stable, all six bonding MOs must be filled and this is possible since each {CH}-unit has three valence electrons.

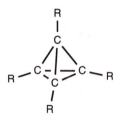

Fig. 4.6 Tetrahedrane.

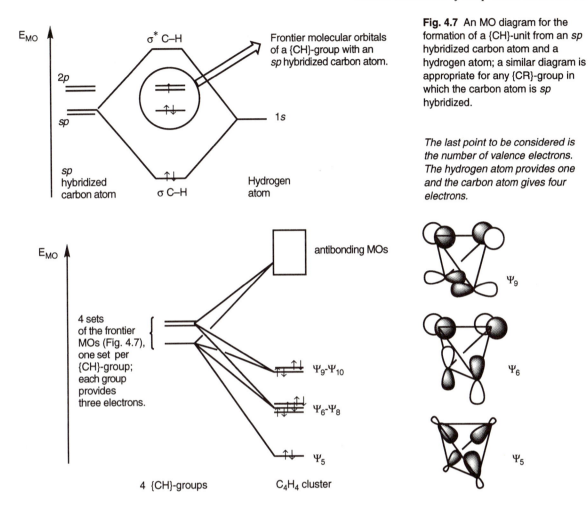

Fig. 4.7 An MO diagram for the formation of a {CH}-unit from an *sp* hybridized carbon atom and a hydrogen atom; a similar diagram is appropriate for any {CR}-group in which the carbon atom is *sp* hybridized.

The last point to be considered is the number of valence electrons. The hydrogen atom provides one and the carbon atom gives four electrons.

Fig. 4.8 A diagram showing the overlap of the frontier orbitals of four {CH}-fragments to form the C_4-core of C_4H_4. The lowest occupied MOs Ψ_{1-4} are the orbitals with terminal C-H bonding character; these are omitted from the diagram.

The question posed in the title of this section was: *is a localized approach to the bonding in small clusters containing carbon sufficient?* The answer is that to a first approximation it is, but if the cluster contains carbon atoms the geometry of which deviates significantly from regular tetrahedral, trigonal planar, or linear, it becomes more appropriate to use a multi-centred bonding description such as described above for C_4H_4.

It is important to appreciate that the number of bonding MOs for the C_4-core of a tetrahedrane C_4R_4 and therefore the number of pairs of cluster bonding electrons is the *same* as the number required in a localized 2-centre-2-electron C–C edge bonding description for the molecule; there are six occupied cluster bonding MOs in C_4R_4 and six C–C edges in the tetrahedron. This is true for other tetrahedral *p*-block clusters, e.g. P_4 and $[Bi_2Sn_2]^{2-}$. Provided that one remembers that it is an *approximate* picture, it is convenient to regard these tetrahedral molecules as possessing localized edge bonding.

4.3 Some clusters of elements in group 15

Each of the tetrahedral clusters E_4 (E = P, As, Sb, or Bi) possesses 20 valence electrons (ve) and, if one lone pair of electrons per atom is localized outside the cluster, there remains the correct number of electrons to form six localized σ-bonds, one per edge of the tetrahedron. The bonding in each of the E_4 clusters is thus related to that in tetrahedrane (see Section 4.2).

20 ve Bi_4
6 pairs of ve for σ-bonding
4 pairs of ve are lone pairs

22 ve $[Bi_4]^{2-}$
4 pairs of ve for σ-bonding plus 6 π-electrons
4 pairs of ve are lone pairs

Fig. 4.9 Cluster opening on going from Bi_4 to $[Bi_4]^{2-}$; (ve = valence electrons).

Bonding   Non-bonding   Non-bonding   Antibonding

Fig. 4.10 π-Molecular orbitals for the square ring of $[Bi_4]^{2-}$.

See Section 4.2 for bonding schemes in tetrahedral clusters.

The addition of two electrons to Bi_4 causes a change in structure from a tetrahedron to an open square (Fig. 4.9). As in Bi_4, one lone pair of electrons is assigned to each bismuth atom in $[Bi_4]^{2-}$. This leaves 14 valence electrons of which eight are used to form four Bi–Bi σ-bonds and six are involved in π-interactions. The molecular orbitals that describe the π-system are shown in Fig. 4.10; only one π-MO possesses ring-bonding character and four of the six electrons occupy non-bonding MOs. The observed Bi–Bi bond lengths of 2.94 Å in $[Bi_4]^{2-}$ indicate little π-character per Bi–Bi edge and this is consistent with the MO description of the bonding. Analogous bonding descriptions are appropriate for the cations $[E_4]^{2+}$ (E = S, Se, or Te) which are isoelectronic with $[Bi_4]^{2-}$.

The pathway for chemical oxidation of P_4 is shown in Fig. 4.11. The distribution of electrons in the P_4 tetrahedron may be described in terms of a localized bonding scheme. Stepwise oxidation occurs first by oxidation of the P–P edges to give localized P–O interactions. Each oxygen atom provides two electrons to each P–P bond in P_4 thereby providing sufficient valence electrons for twelve 2-centre-2-electron P–O interactions in P_4O_6. Complete oxidation to P_4O_{10} involves use of each phosphorus atom lone pair. In keeping with the resonance structures shown at the bottom right of Fig. 4.11, the terminal P–O bond lengths in P_4O_{10} are shorter (1.43 Å) than those associated with the bridging P–O–P interactions (1.60 Å). The phosphorus–phosphorus distance of 2.21 Å in P_4 increases to more than 2.8 Å in the oxide cages. This is consistent with an initially bonding interaction becoming non-bonding; the cluster has expanded in size and the P–O–P interactions now support it. Analogous localized bonding schemes apply to the clusters P_4S_{10}, $P_4O_4S_6$, and $P_4O_6S_4$ (see Fig. 3.36). P_4S_3 is derived from P_4 by having just three P–P edges broken as a result of interactions with sulfur atoms. $[P_7]^{3-}$ and P_7R_3 (R = alkyl) are related to P_4S_3; each S atom is replaced by P^- or a PR-group respectively.

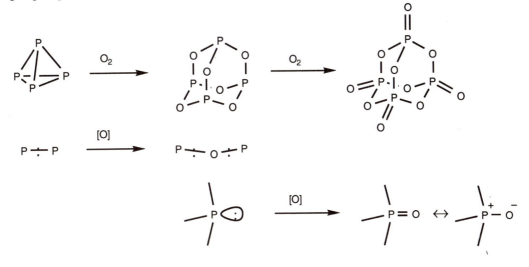

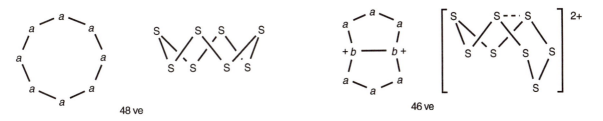

Fig. 4.11   Conversion of P_4 to P_4O_6 and P_4O_{10}.

4.4   Bonding in rings and clusters of group 16 elements

One form of elemental sulfur consists of S_8 rings for which a simple 2-centre-2-electron bonding approach is valid. Of the six valence electrons per S atom, two are used for single bond formation to two adjacent sulfur atoms and four electrons are accommodated as two stereochemically active lone pairs. The ring is a 48 ve system. Upon a two electron oxidation the 46 ve cation $[S_8]^{2+}$ is formed. Oxidation is accompanied by folding of the ring and the formation of a transannular S---S interaction (Fig. 4.12).

ve = valence electrons

48 ve

46 ve

Fig. 4.12   The structural change that accompanies the two electron oxidation of S_8. An *a*-type atom carries two lone pairs of electrons and a *b*-type atom carries one lone pair. The reason for the dotted-line in the final diagram is explained in the text.

The bonding in $[S_8]^{2+}$ may be rationalized if the atoms are separated out into types *a* and *b* according to their connectivity (Fig 4.12). Each 2-coordinate sulfur atom (type *a*) carries two lone pairs as in S_8 whilst each atom involved in the transannular interaction (type *b*) bears only one lone pair. Of the 46 ve, $(\{6 \times 4\} + \{2 \times 2\})$ electrons are present as lone pairs

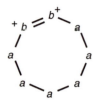

Fig. 4.13 Resonance structure for $[S_8]^{2+}$ without a transannular bond and showing atom types *a* and *b*

leaving 18 ve available for bonding. Thus, the dication will be satisfied if it forms nine covalent bonds; hence the formation of the transannular bond may be rationalized. However, other resonance structures, each without a transannular bond but with an S=S double bond could be drawn (Fig. 4.13) and contributions from these will effectively diminish the importance of the transannular interaction. The observed S–S bond distances in the S_8 and $[S_8]^{2+}$ rings are 2.06 Å and 2.04 Å, respectively, whilst the transannular S--S distance in $[S_8]^{2+}$ is 2.83 Å.

The method outlined above may be applied to the six-atom species $[E_6]^{n+}$ for E = group 16 element. At one end of the series is neutral E_6, exemplified by S_6. A localized bonding description is appropriate here since there are 36 ve and allowing each atom to be an *a*-type atom with two lone pairs provides 6 pairs of electrons for bonding. Thus, E_6 is a ring with six 2-centre-2-electron bonds. At the other extreme, a trigonal prismatic cluster having nine localized edge bonds could be envisaged for $[E_6]^{6+}$ in which each atom is of type *b*. In fact, no hexacation of this type has been isolated. The cation $[Te_6]^{4+}$ is known. $[E_6]^{4+}$ is a 32 ve species and electrons can be distributed such that the cluster consists of four *b*-type and two *a*-type atoms; this gives 16 ve for cluster bonding. There are two end points from which a structure might be predicted. At the top of Fig. 4.14 the oxidized species $[E_6]^{4+}$ is derived from an E_6-ring. Only four of the possible resonance structures for $[E_6]^{4+}$ are drawn. At the bottom of Fig. 4.14, $[E_6]^{4+}$ is derived by reducing the hypothetical $[E_6]^{6+}$; addition of a pair of electrons leads to the cleavage of one E–E bond and three possible resonance structures for the tetracation are shown. The two schemes lead to a common result as indicated in the figure. The experimentally determined structure of $[Te_6]^{4+}$ shows that an elongated trigonal prism is adopted (see Fig. 3.43). This may be rationalized in terms of combining the three resonance structures shown at the bottom of Fig. 4.14. Alternatively, the results of an MO treatment of the bonding in $[E_6]^{4+}$ show that a cluster antibonding orbital (Fig. 4.15) is occupied and this is consistent with the observed elongation of the prism.

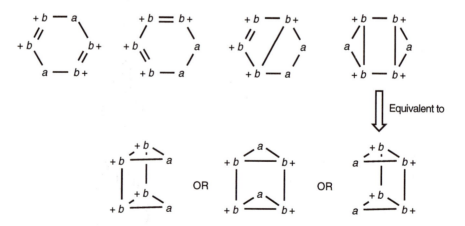

Fig. 4.14   Resonance structures for an $[E_6]^{4+}$ cation. Their generation is considered in terms of the oxidation of an E_6 ring (top of the diagram) and in terms of the reduction of the hypothetical $[E_6]^{6+}$ cation (bottom line of the diagram).

The bonding in the dication $[E_6]^{2+}$ (e.g. $[Te_3S_3]^{2+}$ or $[Te_2Se_4]^{2+}$) may be rationalized by considering either the two electron oxidation of E_6 (bond formation) or the two electron reduction of $[E_6]^{4+}$ (bond breakage). The results of this are shown in Fig. 4.16. Of the 34 ve available in $[E_6]^{2+}$, 14 ve are used to form seven E–E bonds.

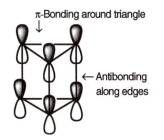

π-Bonding around triangle

← Antibonding along edges

Fig. 4.15 Elongation of the prism of $[E_6]^{4+}$ by occupation of a partially antibonding MO.

4.5 Clusters in which donor–acceptor links are important

In Section 4.1, the importance of coordinate (dative) bond formation between atoms or molecules, one of which is electron rich and the other electron poor, was noted. Intermolecular bond formation is responsible not only for the production of dimers such as Al_2Cl_6, but also for trimers and higher oligomers. The structures of clusters such as $[MeNBCl_2]_4$ (Fig. 3.13), iminoalanes (Figs. 3.14 and 3.5), the cubane $[^iBuAlPSiPh_3]_4$, and thallium-containing cubanes (Fig. 3.17) to name but a few rely upon donor–acceptor interactions.

The bonding scheme is most easily described with a monocyclic example. Consider the formation of the cyclic trimer $[Me_2PBH_2]_3$ (Fig. 4.17). In monomeric Me_2PBH_2, a σ-bonded (2-centre-2-electron) framework leaves a lone pair of electrons on the phosphorus atom and an empty $2p$ AO on the boron atom. Donation of the lone pair from phosphorus to boron could occur intra- or inter-molecularly. The latter leads to oligomerization. Despite the relative complexity of some of the clusters illustrated in, for example, Figs. 3.14 and 3.17, the bonding in each molecule may be rationalized in terms of a localized bonding scheme similar to that in $[Me_2PBH_2]_3$.

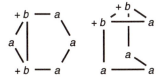

Fig. 4.16 Resonance structures for $[E_6]^{2+}$ viewed in terms of oxidation of an E_6-ring or reduction of an $[E_6]^{4+}$ prism. The structures are equivalent.

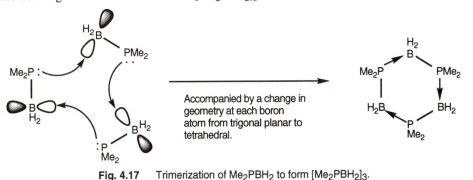

Accompanied by a change in geometry at each boron atom from trigonal planar to tetrahedral.

Fig. 4.17   Trimerization of Me_2PBH_2 to form $[Me_2PBH_2]_3$.

4.6 Bonding in electron deficient clusters

B–H–B bridge bonds in diborane(6)

Before considering electron deficient clusters, it is instructive to look briefly at the bonding in diborane(6), B_2H_6. The boron atom in BH_3 possesses a vacant $2p$ AO which readily accepts a pair of electrons from a Lewis base, L (Fig. 4.18). In the absence of a Lewis base, BH_3 will dimerize by using a pair of B–H bonding electrons to donate into the empty $2p$ AO of a second molecule of BH_3. This is a two-way process, but complete donation would lead simply to hydrogen atom exchange and neither BH_3 molecule would

The term *electron deficient* is defined in Section 1.2.

A B–H–B bridging interaction is an example of a 3-centre-2-electron bond.

gain in terms of filling the vacant orbital. However, if each donation is only partially achieved, the result is that two bonding electrons are shared between three atomic centres and a B–H–B bridge is formed (Fig. 4.18). In this way, each boron atom is effectively satisfied in terms of the octet rule, at least until a more potent Lewis base comes along (see Section 6.2). The formation of 3-centre-2-electron interactions is one way of overcoming electron deficiency.

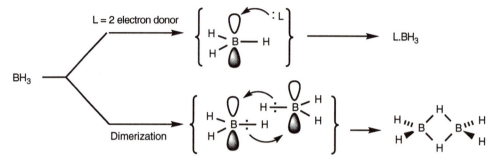

Fig. 4.18 Two ways of achieving a full octet for the boron atom in BH₃

Fig. 4.19 Structure of [Sn₅]²⁻.

A *Zintl ion* is defined in Section 3.7.

Lipscomb's *styx* rules

Cluster molecules of group 13 elements and in particular of the element boron, and to a lesser extent of elements in group 14, exhibit structures with bonding which cannot conveniently be described in terms of localized 2-centre-2-electron interactions. Consider the Zintl ion [Sn₅]²⁻ which adopts a closed trigonal bipyramidal structure (Fig. 4.19). [Sn₅]²⁻ possesses 22 ve and with one lone pair per atom, only 12 ve remain for cluster bonding. This is clearly an insufficient number for localization along the nine Sn–Sn edges in the cage. Similarly, in B₅H₉ (Fig. 4.20) the connectivities of the boron and of four of the hydrogen atoms surpass the number of valence electrons available per atom.

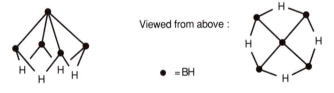

Fig. 4.20 Structure of B₅H₉.

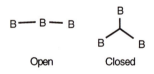

Fig. 4.21 B–B–B interactions may be open or closed.

The distribution of valence electrons in borane clusters may be viewed in terms of a combination of localized 2-centre- and 3-centre-2-electron interactions. This method was developed by Lipscomb and is defined in a set of equations known as *styx* rules. An appropriate bonding scheme for the molecule is derived by solving the simultaneous equations given in Eqn 4.2 where *s* is the number of 3-centre-2-electron B–H–B interactions, *t* is the number of 3-centre-2-electron B–B–B interactions (Fig. 4.21), *y* is the number of 2-centre-2-electron B–B bonds, and x is the number of BH$_2$-units. The formula of the borane is given by B_pH_{p+q}.

$$s+x=q \qquad s+t=p \qquad t+y=p-{}^q/_2 \qquad \text{Eqn 4.2}$$

An exercise in *styx* rules exemplified by B$_5$H$_9$:

1. Consider a particular borane and define p and q:
   For B$_5$H$_9$: $p=5$ $q=4$.
2. Choose integral values of s between the limits $0 \le s \le p$:
   For B$_5$H$_9$: $0 \le s \le 5$.
3. Solve Eqn 4.2 for each value of s:

   For $s=0$, $x=4$ $t=5$ $y=-2$
   For $s=1$, $x=3$ $t=4$ $y=-1$
   For $s=2$, $x=2$ $t=3$ $y=0$
   For $s=3$, $x=1$ $t=2$ $y=1$
   For $s=4$, $x=0$ $t=1$ $y=2$
   For $s=5$, $x=-1$ $t=0$ $y=3$

4. Discard solutions from (3.) which include negative values of x, t, or y as these are not real; (see the definitions of *styx* which require positive values):
   For B$_5$H$_9$, only three solutions remain:
   styx = (2302)   or   (3211)   or   (4120)
5. Interpret the real solutions of *styx* in terms of structural diagrams:
   The possible structures deduced for B$_5$H$_9$ are shown in Fig. 4.22. For example, the result *styx* = (2302) means that there are two 3-centre-2-electron B–H–B interactions, three 3-centre-2-electron B–B–B interactions, no 2-centre-2-electron B–B bonds, and two BH$_2$-units. Although it is difficult to choose between the predicted forms, in hindsight, the experimentally determined structure implies that (4120) is preferred.

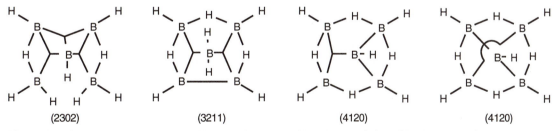

(2302)   (3211)   (4120)   (4120)

Fig. 4.22   Structures for B$_5$H$_9$ deduced by use of *styx* rules. Note that *styx* = (4120) leads to two valence structures.

Polyhedral Skeletal Electron Pair Theory

For a relatively small cluster such as B$_5$H$_9$, the use of *styx* rules is fairly straightforward and a localized bonding scheme can be developed. However, the method does not allow a structure to be unambiguously assigned. A second approach to the problem of bonding in cluster molecules has been provided by Wade, Williams, and Mingos in the form of the *Polyhedral Skeletal Electron Pair Theory* (*PSEPT*), also known as Wade's rules. The method gives an empirical set of rules (see below) which may be used either to deduce a structure for a given chemical formula or to rationalize an experimentally determined cluster structure.

PSEPT is based on the fact that a closed deltahedral cluster core with n vertices requires $(n + 1)$ pairs of electrons to fill $(n + 1)$ skeletal bonding MOs. An MO justfication for this is shown on the following page but for the purposes of this text, an empirical treatment will be used.

PSEPT may be summarized as follows:

1. A *closo*-deltahedral cluster core with n vertices requires $(n + 1)$ pairs of electrons to occupy $(n + 1)$ cluster bonding MOs.

2. If the parent (closed) deltahedron has n vertices, then the related *nido*-cluster has $(n - 1)$ vertices and $(n + 1)$ pairs of electrons to occupy $(n + 1)$ cluster bonding MOs, the related *arachno*-cluster has $(n - 2)$ vertices and $(n + 1)$ pairs of electrons to occupy $(n + 1)$ cluster bonding MOs, and the related *hypho*-cluster has $(n - 3)$ vertices and $(n + 1)$ pairs of electrons to occupy $(n + 1)$ cluster bonding MOs.

In *PSEPT*, clusters are classed according to their relationship to a closed deltahedral cage. A cluster with a closed deltahedral structure is called a *closo*-cluster. If it lacks one vertex with respect to the parent skeleton, it is a *nido*-cluster. If it has two vertices missing, it is an *arachno*-cluster. If there are three vertices vacant, it is a *hypho*-cluster (Fig. 4.23).

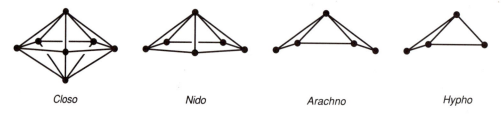

| Closo | Nido | Arachno | Hypho |

Fig. 4.23 The relationship between *closo*-, *nido*-, *arachno*-, and *hypho*-clusters exemplified by the pentagonal bipyramidal skeleton.

PSEPT applied to Zintl ions

The empirical rules which constitute *PSEPT* are demonstrated below with examples from Zintl ion chemistry. In each case, the vertex atom carries a lone pair of electrons and the remaining valence electrons are used for cluster bonding. Examples 1 and 2 begin with known structures and derive the class of cluster. Example 3 predicts a structure from a given formula.

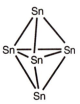

Example 1: $[Sn_5]^{2-}$

Rationalize why the structure of $[Sn_5]^{2-}$ is a closed trigonal bipyramid.

Sn is in group 14.

Each Sn atom carries a lone pair and gives 2 ve to cluster bonding.

There are an additional 2 ve from the dinegative charge.

$$\text{Total electrons available} = (5 \times 2) + 2$$
$$= 12 \text{ ve}$$
$$= 6 \text{ pairs}$$

$\therefore$ $[Sn_5]^{2-}$ has 6 pairs of electrons with which to bond 5 cluster atoms.

$\therefore$ The requirement for a *closo*-trigonal bipyramidal structure is met.

Molecular orbital treatment for the cluster dianion [B₆H₆]²⁻

The dianion $[B_6H_6]^{2-}$ is an octahedral cluster of six boron atoms each of which carries a terminal hydrogen atom. Just as for carbon in a tetrahedrane cluster, each boron atom in $[B_6H_6]^{2-}$ may be considered to be *sp* hybridized. One *sp* hybrid is used to bond to the terminal hydrogen atom and one is available for cluster bonding. The frontier MOs of each {BH}-unit are equivalent to those of the {CH} unit shown in Fig. 4.7; only the number of valence electrons is different.

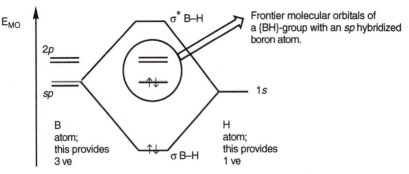

Frontier molecular orbitals of a {BH}-group with an *sp* hybridized boron atom.

B atom; this provides 3 ve

H atom; this provides 1 ve

Since each {BH}-unit provides three frontier orbitals, the total number of MOs expected for the cluster core of $[B_6H_6]^{2-}$ is (6 x 3) = 18. The frontier MOs of the six {BH}-units combine to give seven bonding MOs and eleven antibonding MOs. The cluster bonding MOs are generated by considering the symmetry of the octahedron and of the orbitals available.

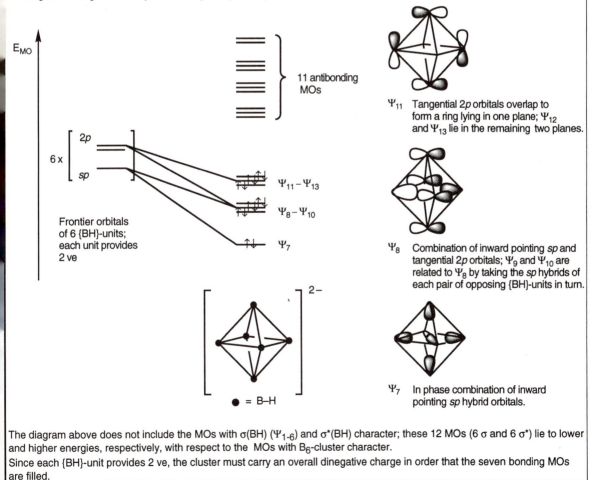

11 antibonding MOs

Frontier orbitals of 6 {BH}-units; each unit provides 2 ve

$\Psi_{11}-\Psi_{13}$

$\Psi_8-\Psi_{10}$

Ψ_7

● = B–H

Ψ_{11} Tangential 2*p* orbitals overlap to form a ring lying in one plane; Ψ_{12} and Ψ_{13} lie in the remaining two planes.

Ψ_8 Combination of inward pointing *sp* and tangential 2*p* orbitals; Ψ_9 and Ψ_{10} are related to Ψ_8 by taking the *sp* hybrids of each pair of opposing {BH}-units in turn.

Ψ_7 In phase combination of inward pointing *sp* hybrid orbitals.

The diagram above does not include the MOs with σ(BH) (Ψ_{1-6}) and σ*(BH) character; these 12 MOs (6 σ and 6 σ*) lie to lower and higher energies, respectively, with respect to the MOs with B_6-cluster character.

Since each {BH}-unit provides 2 ve, the cluster must carry an overall dinegative charge in order that the seven bonding MOs are filled.

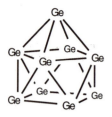

Example 2: [Ge₉]⁴⁻

Rationalize why [Ge₉]⁴⁻ adopts a monocapped square antiprismatic cage.

Ge is in group 14.

Each Ge atom carries a lone pair and gives 2 ve to cluster bonding.

There are an additional 4 ve from the tetranegative charge.

$$
\begin{aligned}
\text{Total electrons available} &= (9 \times 2) + 4 \\
&= 22 \text{ ve} \\
&= 11 \text{ pairs}
\end{aligned}
$$

∴ The parent deltahedron upon which the structure of $[Ge_9]^{4-}$ is based has 10 vertices and is a bicapped square antiprism. One vertex must be removed to generate a cluster suitable for the 9 germanium atoms in $[Ge_9]^{4-}$.

∴ The requirement for a *nido*-cage, a monocapped square antiprism, is met.

Removal of vertices from a deltahedral framework

The first vertex to be removed tends to be the one of highest connectivity (e.g. as for the pentagonal bipyramid in Fig. 4.23) but in the case of a capped structure (e.g. the tricapped trigonal prism or bicapped square antiprism), one capping-vertex is removed first.

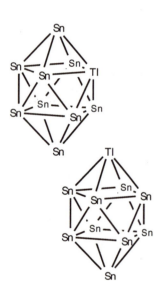

Example 3: [TlSn₉]³⁻

Predict the structure of [TlSn₉]³⁻.

Tl is in group 13 and Sn is in group 14.

Each Tl atom carries a lone pair and gives 1 ve to cluster bonding.

Each Sn atom carries a lone pair and gives 2 ve to cluster bonding.

There are an additional 3 ve from the trinegative charge.

$$
\begin{aligned}
\text{Total electrons available} &= (1 \times 1) + (9 \times 2) + 3 \\
&= 22 \text{ ve} \\
&= 11 \text{ pairs}
\end{aligned}
$$

∴ $[TlSn_9]^{3-}$ has 11 pairs of electrons with which to bond 10 cluster atoms.

∴ There are $(n + 1)$ pairs of electrons for n atoms.

∴ It is predicted that $[TlSn_9]^{3-}$ will adopt a *closo*-bicapped square antiprismatic structure. However, the bicapped square antiprism has two distinct sites. Each capping-site has a connectivity of four and each site within the prism has a connectivity of five. Thus two isomers of $[TlSn_9]^{3-}$ are predicted with the thallium atom in one of two sites. The experimental structure of $[TlSn_9]^{3-}$ shows that the thallium atom prefers to occupy a capping (apical) site.

PSEPT applied to borane and carbaborane clusters

Most of the structures of borane and carbaborane clusters may be rationalized in terms of *PSEPT*. In addition to the vertex {BH}- or {CH}-units, additional hydrogen atoms may have to be accommodated. Each {BH}-unit provides 2 ve to cluster bonding since two electrons are used to form the localized 2-centre-2-electron terminal BH bond (see p.52). Similarly, a {CH}-unit provides 3 ve. Each additional hydrogen atom, be it *terminal or bridging*, supplies one electron to cluster bonding.

A set of guidelines may be followed in order to predict the structure of a borane cluster

1. How many {BH}-units are there?
2. How many additional H atoms are there?
3. How many valence electrons are available for cluster bonding?
4. What is the parent deltahedron?
5. After each {BH}-unit has been accommodated at a skeletal vertex, are there any vertices vacant in the parent deltahedron? What class is the cluster?
6. Additional H atoms are placed either around the B–B edges of an open face of the cluster or in extra terminal positions usually if a boron atom is of particularly low connectivity. Cluster symmetry is generally kept as high as possible.

Examples of structure rationalization and prediction in borane and carbaborane chemistry are given below.

Example 1: $[B_{12}H_{12}]^{2-}$

Rationalize why the structure of $[B_{12}H_{12}]^{2-}$ is an icosahedral cage.

There are 12 {BH}-units and no additional H atoms.

Each {BH}-unit provides 2 ve to cluster bonding.

There are 2 additional electrons from the dinegative charge.

Total electrons available $= (12 \times 2) + 2$

$\qquad\qquad\qquad = 26$ ve

$\qquad\qquad\qquad = 13$ pairs

∴ $[B_{12}H_{12}]^{2-}$ has 13 pairs of electrons with which to bond 12 {BH}-units.

∴ There are $(n + 1)$ pairs of electrons for n units.

∴ $[B_{12}H_{12}]^{2-}$ is a *closo*-cluster with a 12-vertex deltahedral cage, i.e. the icosahedron is adopted.

● = BH

Example 2: B_5H_9

Rationalize why the structure of B_5H_9 is based on a square based pyramid. What is the predicted arrangement of the hydrogen atoms?

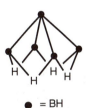

There are 5 {BH}-units and 4 additional H atoms.

Each {BH}-unit provides 2 ve to cluster bonding.

Each extra H atom provides 1 ve to cluster bonding.

Total electrons available $= (5 \times 2) + (4 \times 1)$

$\qquad\qquad\qquad = 14$ ve

$\qquad\qquad\qquad = 7$ pairs

● = BH

∴ B_5H_9 has 7 pairs of electrons with which to bond 5 {BH}-units.

7 pairs of electrons are consistent with a 6-vertex parent deltahedron, i.e. the octahedron.

∴ The B_5-core of B_5H_9 will be based on an octahedron with one vertex vacant, i.e. a square based pyramid.

The four additional hydrogen atoms are accommodated in B–H–B bridging positions around the open face of the pyramid.

Why do bridging hydrogen atoms occupy sites around the open faces of a cluster?

The open face of a cluster has been generated by removing one or more vertices from a deltahedral cluster. From an MO viewpoint, the loss of a vertex leaves a region of electron density focused around the open face. Hence, protons will readily interact with the open face:

H 1s
AO

Occupied cluster MO derived from $[B_6H_6]^{2-}$.

This leads to B–H–B bridges around the open face.

Example 3: $[B_6H_9]^-$

Predict the structure of $[B_6H_9]^-$

There are 6 {BH}-units and 3 additional H atoms.
Each {BH}-unit provides 2 ve to cluster bonding.
Each extra H atom provides 1 ve to cluster bonding.
There is an additional electron from the charge.

$$\text{Total electrons available} = (6 \times 2) + (3 \times 1) + 1$$
$$= 16 \text{ ve}$$
$$= 8 \text{ pairs}$$

∴ $[B_6H_9]^-$ has 8 pairs of electrons with which to bond 6 {BH}-units.

8 pairs of electrons are consistent with a 7-vertex parent deltahedron, i.e. a pentagonal bipyramid.

∴ The B_6-core of $[B_6H_9]^-$ will be based on a pentagonal bipyramid with one vertex vacant, i.e. a pentagonal pyramid

Remove
a vertex

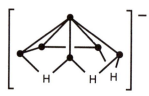

The three extra H atoms form B–H–B bridges along three of the five edges of the open face of the cluster. The B_6-core of $[B_6H_9]^-$ cluster possesses a five-fold rotation axis of symmetry. The presence of the three B–H–B bridges reduces the molecular symmetry. In solution, it is likely that the anion will undergo a fluxional process in which the three bridging hydrogen atoms migrate around the open face of the cluster. In this way, $[B_6H_9]^-$ retains a five-fold axis of symmetry on the NMR spectroscopic timescale.

The guidelines for borane structure prediction may be adapted for carbaborane clusters remembering the following:

1. Carbon atoms tend to be apart from one another in a cluster; B–C interactions are maximized.
2. B–H–B bridging interactions are preferred over C–H–B or C–H–C interactions; there is little evidence for C-H-B bridges and none for C–H–C interactions in carbaborane clusters.

Example 4: $C_2B_4H_6$

Predict the structure of $C_2B_4H_6$.

There are 4 {BH}-units and 2 {CH}-units.

Each {BH}-unit provides 2 ve to cluster bonding.

Each {CH}-unit provides 3 ve to cluster bonding.

$$\text{Total electrons available} = (4 \times 2) + (2 \times 3)$$
$$= 14 \text{ ve}$$
$$= 7 \text{ pairs}$$

∴ $C_2B_4H_6$ has 7 pairs of electrons with which to bond 6 skeletal units.

7 pairs of electrons are consistent with a 6-vertex parent deltahedron, i.e. the octahedron, and $C_2B_4H_6$ will be a *closo*-cluster.

Two isomers are possible, either 1,2-$C_2B_4H_6$ or 1,6-$C_2B_4H_6$.

1,2-isomer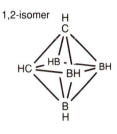

1,6-isomer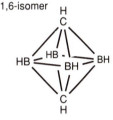

The isolobal principle

It is possible to convert a borane into a carbaborane cluster by replacing a {BH⁻} by a {CH}-unit because both fragments possess the same frontier orbital properties. The two sets of MOs have the same symmetry characteristics, are of approximately the same energy, and contain the same number of electrons available for cluster bonding. The {BH⁻} and {CH} fragments are said to be *isolobal*. The principle can be extended to include a range of atoms and molecular fragments all of which exhibit similar frontier orbitals. Series of isolobal *p*-block fragments are as follows:

{BH⁻} ≡ {BR⁻} ≡ {CH} ≡ {CR} ≡ {NR⁺}   where R is a one electron donor e.g. alkyl, aryl, halogen atom

{BR} ≡ {AlR} ≡ {GaR} ≡ {GeR⁺} ≡ {SR³⁺}   where R is a one electron donor e.g. alkyl, aryl, halogen atom

{BH} ≡ {Tl⁻} ≡ {Sn} ≡ {Pb} ≡ {N⁺} ≡ {S²⁺}   where each bare atom carries an *exo*-lone pair of electrons

The isolobal principle is not restricted to *p*-block elements. Certain transition metal fragments are isolobal with a {BH}-fragment and one of these is the conical (C_{3v} symmetry) {M(CO)₃}-unit where M is a group 8 metal (Fe, Ru, or Os). The frontier MOs of the {M(CO)₃} unit are drawn in Fig. 4.24. The same diagram is appropriate for C_{3v} {M'(CO)₃⁺}-units (M' = Co, Rh, or Ir), or C_{3v} {M"(CO)₃⁻}-units (M" = Mn or Re) units.

The similarities in orbital properties between a {BH}- and C_{3v} {M(CO)₃}-fragment (M = group 8 metal) allows a {BH}-unit in a borane cluster to be replaced by an {M(CO)₃}-unit. This leads to a series of *metallaborane* clusters. When applying *PSEPT* to a metallaborane compound containing an {M(CO)₃}-unit (M = group 8 metal), the unit is treated as a source of two valence electrons. Moving to the left or right of group 8 in the periodic table

The orientation of the ligands attached to the metal atom affects the number and symmetries of the frontier MOs. An {M(CO)₃}-unit is isolobal with a {BH}-fragment if the three carbonyl ligands define a cone so that the fragment has C_{3v} symmetry.

decreases or increases, respectively, the number of electrons provided by a transition metal tricarbonyl unit. Since different *exo*-ligands donate different numbers of electrons to the metal atom, the number of valence electrons provided by a transition metal fragment for cluster bonding can also be adjusted by altering the nature of the ligands attached to the metal as indicated in Eqn 4.3.

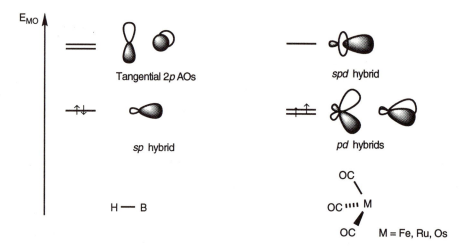

Fig. 4.24 Comparison of the three frontier MOs of {BH}- and C_{3v} {M(CO)$_3$}-fragments (M = group 8 metal). The fragments are isolobal because the number, symmetries, and approximate energies of the orbitals are equal, and each set contains the same number of electrons. The *order* of the MOs does not influence the bonding capability of the fragment.

If x = number of valence electrons provided by a transition metal fragment for cluster bonding, then:

$$x = v + n - 12 \qquad \textbf{Eqn 4.3}$$

where v = number of ve of the metal atom and n = number of electrons donated by the ligands.

CO is a two electron donor ligand.

PR$_3$ is a two electron donor ligand.

An η^n-C$_n$H$_n$ ring is an n π-electron donor ligand.

NO may be a one or three electron donor ligand.

Application of Eqn 4.3 shows that the cluster unit {Os(CO)$_3$} provides 2 ve since $x = (8 + 6 - 12)$. Similarly, {Co(PPh$_3$)$_2$} provides 1 ve since $x = (9 + 4 - 12)$, {Ni(η^5-C$_5$H$_5$)} provides 3 ve since $x = (10 + 5 - 12)$, and {Rh(PPh$_3$)(CO)$_2$} provides 3 ve since $x = (9 + 6 - 12)$.

The isolobal principle combined with *PSEPT* suggests that it should be possible to systematically replace {BH}-units in a borane cluster by, for example, {Ru(CO)$_3$}-units. Thus, starting from B$_5$H$_9$, a series of metallaboranes including B$_4$H$_8$Ru(CO)$_3$, B$_3$H$_7$Ru$_2$(CO)$_6$, and B$_2$H$_6$Ru$_3$(CO)$_9$ can be postulated. References detailing metallaboranes are listed in the further reading list (Section 1.4).

5   Synthetic routes to clusters

5.1   Boranes and hydroborate anions

Precursors

The precursors in many syntheses of boranes and their anionic derivatives
are diborane(6) and the octahydrotriborate(1–) anion. B_2H_6 is a gas under
atmospheric conditions (b.p. = –92.5°C). It may be prepared according to
Eqn 5.1 from commercially available starting materials. A convenient *in situ*
preparation is given in Eqn 5.2. Adducts $L.BH_3$ may be prepared or obtained
commercially, e.g. $THF.BH_3$ (Fig. 5.1).

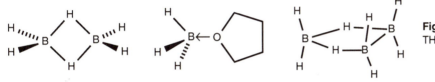

Fig. 5.1   Structures of B_2H_6, $THF.BH_3$, and $[B_3H_8]^-$.

$$2\ Na[BH_4] + I_2 \xrightarrow[\text{25°C}]{\text{diglyme}} B_2H_6 + 2\ NaI + H_2 \qquad \textbf{Eqn 5.1}$$

$$3\ Na[BH_4] + 4\ Et_2O.BF_3 \xrightarrow[\text{25°C}]{\text{diglyme}} 2\ B_2H_6 + 3\ Na[BF_4] + 4\ Et_2O \qquad \textbf{Eqn 5.2}$$

$$3\ Na[BH_4] + I_2 \xrightarrow[\text{100°C}]{\text{diglyme}} 2\ NaI + 2\ H_2 + Na[B_3H_8] \qquad \textbf{Eqn 5.3}$$

$$5\ Na[BH_4] + 4\ Et_2O.BF_3 \xrightarrow[\text{100°C}]{\text{diglyme}} 3\ Na[BF_4] + 2\ Na[B_3H_8] + 2\ H_2 + 4\ Et_2O \qquad \textbf{Eqn 5.4}$$

The structure of the octahydrotriborate(1–) ion is shown in Fig. 5.1 and
methods of preparation are given in Eqns 5.3 and 5.4. Alkali metal or
tetraalkyl ammonium salts of $[B_3H_8]^-$ are white, crystalline solids.

In solution, the $[B_3H_8]^-$ ion is highly fluxional with all eight hydrogen
atoms (and all three boron atoms) being equivalent on the NMR
spectroscopic timescale even at low temperatures. The ^{11}B and ^1H NMR
spectral resonances are a binomial nonet and a non-binomial 10-line
multiplet, respectively.

$[B_nH_n]^{2-}$ dianions

Two general methods for the synthesis of the *closo*-$[B_nH_n]^{2-}$ dianions are
given in Eqns 5.5 and 5.6. The first involves the thermolysis of a neutral
borane in the presence of an adduct $L.BH_3$ in which L may be an amine or
hydride ion. In the second route a Lewis base initially interacts with the
borane displacing molecular hydrogen; proton transfer from borane to Lewis
base then generates the *closo*-cluster.

Multiplicity of NMR signal = $2nI + 1$
I = nuclear spin
n = number of equivalent nuclei
   coupling to observed nucleus.
For ^{11}B $I = \frac{3}{2}$ and ^1H $I = \frac{1}{2}$.
For $[B_3H_8]^-$:
For ^1H NMR observed:
   Multiplicity = $2(3)(\frac{3}{2}) + 1 = 10$
For ^{11}B NMR observed:
   Multiplicity = $2(8)(\frac{1}{2}) + 1 = 9$

$$2\ Et_3N.BH_3 + B_{10}H_{14} \xrightarrow{90°C} [Et_3NH]_2[B_{12}H_{12}] + 3\ H_2 \qquad \text{Eqn 5.5}$$

$$2\ Et_3N + B_{10}H_{14} \xrightarrow[-H_2]{} \{B_{10}H_{12}(NEt_3)_2\} \longrightarrow [Et_3NH]_2[B_{10}H_{10}] \qquad \text{Eqn 5.6}$$

Routes for the synthesis of specific *closo*-$[B_nH_n]^{2-}$ anions are shown in Eqns 5.7 and 5.8. The anions $[B_7H_7]^{2-}$ and $[B_8H_8]^{2-}$ are obtained by the aerial degradation of $[B_9H_9]^{2-}$, itself obtained in a mixture with $[B_{10}H_{10}]^{2-}$ and $[B_{12}H_{12}]^{2-}$ from the thermolysis at 230°C of $Cs[B_3H_8]$.

$$2\ Na[B_3H_8] \xrightarrow[\text{diglyme}]{160°C} Na_2[B_6H_6] + 5\ H_2 \qquad \text{Eqn 5.7}$$

$$2\ Cs_2[B_{11}H_{13}] \xrightarrow[-H_2]{250°C} 2\ Cs_2[B_{11}H_{11}] \xrightarrow{600°C} Cs_2[B_{10}H_{10}] + Cs_2[B_{12}H_{12}] \qquad \text{Eqn 5.8}$$

Single cage *nido*- and *arachno*-boranes

A *hot–cold reactor* is designed to give an interface between two regions of extreme temperatures. A vessel containing electrically heated oil (temperature T_1) is surrounded by a cold bath (temperature T_2). A tube carries B_2H_6 through the hot-cold interface where decomposition occurs. The products, controlled by varying T_1 and T_2, are led out through the cold region.

Several *nido*- and *arachno*-boranes are prepared by the pyrolysis of diborane(6) in the vapour phase in a hot–cold reactor with a temperature interface of T_1–T_2. With $T_1 = 120°C$ and $T_2 = -78°C$ or $-30°C$, B_4H_{10} or B_5H_{11}, respectively, is formed. B_5H_9 is produced with $T_1 = 180°C$ and $T_2 = -78°C$. Heating B_2H_6 at 180–220°C under static conditions yields $B_{10}H_{14}$.

Salts of the octahydrotriborate(1−) anion are more convenient starting materials than diborane(6). Protonation of $[B_3H_8]^-$ produces not B_3H_9 (which is not a stable species) but B_4H_{10} (Eqn 5.9) or B_5H_{11}. Abstraction of H^- from $[B_3H_8]^-$ by a Lewis acid generates the unstable $\{B_3H_7\}$ which spontaneously adds a $\{BH_3\}$-unit to give B_4H_{10}. Related syntheses are shown in Eqns 5.10 and 5.11.

$$4\ Na[B_3H_8] + 4\ HCl \longrightarrow 3\ B_4H_{10} + 3\ H_2 + 4\ NaCl \qquad \text{Eqn 5.9}$$

$$[B_4H_9]^- + BCl_3 \xrightarrow{-35°C} B_5H_{11} + [HBCl_3]^- + \text{Polymeric products} \qquad \text{Eqn 5.10}$$

$$[B_9H_{14}]^- + BCl_3 \xrightarrow{25°C} B_{10}H_{14} + H_2 + [HBCl_3]^- + \text{Polymeric products} \qquad \text{Eqn 5.11}$$

Cluster build-up via a halogenated derivative is exemplified in synthetic routes to the two *nido*-clusters B_5H_9 and B_6H_{10} (Eqns 5.12 and 5.13). The related *arachno*-boranes B_5H_{11} and B_6H_{12} can be synthesized via the reaction of diborane(6) (which acts as a source of a BH_3-unit) with $[B_4H_9]^-$ and $[B_5H_8]^-$, respectively, (Eqns 5.14 and 5.15). Note that although the two precursors are *arachno*- and *nido*-cluster anions, respectively, both products are *arachno*-species since protonation of the intermediate anion $[B_5H_{12}]^-$ in Eqn 5.14 is accompanied by dihydrogen loss.

KH (Eqn 5.13) and NaH are deprotonating agents, removing bridging rather than terminal hydrogen atoms from a borane cluster (see Fig. 6.4).

$$5\ [B_3H_8]^- + 5\ HBr \xrightarrow[-H_2]{} 5\ [B_3H_7Br]^- \xrightarrow{100°C} 3\ B_5H_9 + 4\ H_2 + 5\ Br^- \qquad \text{Eqn 5.12}$$

$$B_5H_9 + Br_2 \xrightarrow[-HBr]{25°C} 1\text{-}BrB_5H_8 \xrightarrow[-H_2]{KH\ -78°C} K[1\text{-}BrB_5H_7] \xrightarrow{B_2H_6\ -78°C} B_6H_{10} + KBr \qquad \text{Eqn 5.13}$$

A rational route to *arachno*-B_5H_{11} is the reduction of its parent *nido*-cluster, B_5H_9 (Eqn 5.16). Adding two electrons to *nido*-B_5H_9 causes the B_5-skeleton to open up (see Section 4.6). Protonation generates a neutral *arachno*-cluster which retains the B_5-framework of the intermediate dianion.

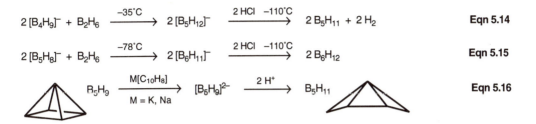

$$2\,[B_4H_9]^- + B_2H_6 \xrightarrow{-35°C} 2\,[B_5H_{12}]^- \xrightarrow{2\,HCl\ \ -110°C} 2\,B_5H_{11} + 2\,H_2 \qquad \textbf{Eqn 5.14}$$

$$2\,[B_5H_8]^- + B_2H_6 \xrightarrow{-78°C} 2\,[B_6H_{11}]^- \xrightarrow{2\,HCl\ \ -110°C} 2\,B_6H_{12} \qquad \textbf{Eqn 5.15}$$

$$B_5H_9 \xrightarrow[M = K,\ Na]{M[C_{10}H_8]} [B_5H_9]^{2-} \xrightarrow{2\,H^+} B_5H_{11} \qquad \textbf{Eqn 5.16}$$

One method of approaching the synthesis of large borane clusters is to oxidatively fuse together two smaller cages. The reactions are promoted by a variety of transition metal compounds such as iron(III) or ruthenium(III) complexes which are reduced during the reactions to iron(II) or ruthenium(II) species, respectively. Choice of oxidizing agent may be critical as shown by the products isolated from the oxidative fusion of two $[B_5H_8]^-$ anions (Eqns 5.17 and 5.18). Use of the deuterium labelled starting material $[1\text{-}DB_5H_7]^-$ provides an insight into the mechanism for cluster fusion (Fig. 5.2).

The term *fusion* is used here to imply that two borane cages react together to produce a new *single* cluster. The term *coupling* is used to mean that two single cages react together to give a product in which the original cages are coupled together by a vertex, an *exo*-B–B bond, or a B–B edge.

$$4\,[B_5H_8]^- \xrightarrow{RuCl_3\ \ THF} B_{10}H_{14} + 2\,B_5H_9 \qquad \textbf{Eqn 5.17}$$

$$4\,[B_5H_8]^- \xrightarrow{FeCl_3\ \ THF} B_{10}H_{14} + 2,2'\text{-}\{B_5H_8\}_2 + H_2 \qquad \textbf{Eqn 5.18}$$

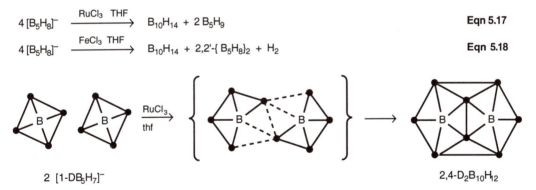

2 $[1\text{-}DB_5H_7]^-$ $2,4\text{-}D_2B_{10}H_{12}$

Fig. 5.2 Stereospecific metal promoted oxidative fusion of two $[1\text{-}DB_5H_7]^-$ clusters to give $2,4\text{-}D_2B_{10}H_{12}$. The sites marked B are the deuterium labelled sites. The cluster structures are drawn as plan-views with only the B_n-skeletons shown.

Coupled borane clusters

As distinct from the fusion of two borane clusters, cluster coupling to give products such as $\{B_5H_8\}_2$ and $\{B_5H_8\}\{B_2H_5\}$ has been achieved by photolysis (Eqns 5.19 and 5.20) and by platinum(II) catalysis (Eqns 5.21–5.23). Note the non-specificity of the photolytic method as opposed to the specific nature of the metal promoted syntheses.

See Figs. 3.2 and 3.4 for atom numbering in B_5- and B_{10}-cages.

$$2\,B_5H_9 \xrightarrow{h\nu} 1,1'\text{-},\ 1,2'\text{-},\ \text{and}\ 2,2'\text{-}\{B_5H_8\}_2 + H_2 \qquad \textbf{Eqn 5.19}$$

$$2\,B_{10}H_{14} \xrightarrow{h\nu} 1,2'\text{- and}\ 2,2'\text{-}\{B_{10}H_{13}\}_2 + H_2 \qquad \textbf{Eqn 5.20}$$

$$2\,B_4H_{10} \xrightarrow{PtBr_2} 1,1'\text{-}\{B_4H_9\}_2 + H_2 \qquad \textbf{Eqn 5.21}$$

$$2\,B_5H_9 \xrightarrow{PtBr_2} 1,2'\text{-}\{B_5H_8\}_2 + H_2 \qquad \textbf{Eqn 5.22}$$

$$B_5H_9 + B_2H_6 \xrightarrow{PtBr_2} \{B_5H_8\}\{B_2H_5\} + H_2 \qquad \textbf{Eqn 5.23}$$

5.2 Carbaborane clusters

In carbaborane clusters, carbon atoms tend:
1. to be apart from one another and
2. to occupy sites of low connectivity.

The area of carbaborane clusters, and hence their synthesis, is a large one and the coverage here is necessarily selective. Examples here are chosen for their general importance or else illustrate a particular strategy of cluster synthesis.

A general method of synthesis, particularly for small carbaborane clusters, is the pyrolysis of a borane with an alkyne. This leads to the elimination of gaseous dihydrogen and the coupling together of the B_n- and C_2-fragments. Such reactions give mixtures of products as shown in Eqn 5.24. Although the carbon atoms are bonded by a carbon–carbon triple bond in the alkyne molecule, the thermodynamically favoured isomers of the carbaborane products formed at elevated temperatures exhibit remote carbon vertices. For example, in $C_2B_4H_6$, the kinetic product is $1,2\text{-}C_2B_4H_6$ but the thermodynamic product is $1,6\text{-}C_2B_4H_6$. There is a trend to maximize B–C interactions and in addition, carbon atoms tend to occupy sites of *low connectivity*. Hence, $2,4\text{-}C_2B_5H_7$ (Eqn 5.24) is preferred over both $1,7\text{-}C_2B_5H_7$ (in which the carbon atoms are apart but are in sites of relatively high connectivity) and $2,3\text{-}C_2B_5H_7$ (where the carbon atoms are in sites of low connectivity but are adjacent to one another). High temperatures favour *closo*-carbaborane clusters (Eqn 5.24). Reactions at lower temperatures may yield *nido*-carbaboranes, e.g. the pyrolysis of pentaborane(9) with acetylene at 200°C gives *nido*-$2,3\text{-}C_2B_4H_8$.

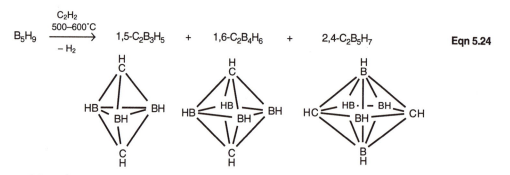

$$B_5H_9 \xrightarrow[- H_2]{\substack{C_2H_2 \\ 500\text{–}600°C}} 1,5\text{-}C_2B_3H_5 \quad + \quad 1,6\text{-}C_2B_4H_6 \quad + \quad 2,4\text{-}C_2B_5H_7 \qquad \textbf{Eqn 5.24}$$

The older nomenclature of *o*-, *m*-, and *p*-$C_2B_{10}H_{12}$ referring to the 1,2-, 1,7- and 1,12-isomers, respectively, is still commonly used.

A general synthesis of $1,2\text{-}C_2B_{10}H_{12}$ is shown in Eqn 5.25. On heating to 470°C, $1,2\text{-}C_2B_{10}H_{12}$ isomerizes to $1,7\text{-}C_2B_{10}H_{12}$, and at 700°C, the 1,12-isomer is obtained.

$$B_{10}H_{14} + 2\,L \xrightarrow{- H_2} B_{10}H_{12}L_2 \xrightarrow{HC \equiv CH} 1,2\text{-}C_2B_{10}H_{12} + 2\,L + H_2$$

$$L = MeCN,$$
$$R_3N,$$
$$R_2S \quad R = alkyl$$

Eqn 5.25

Treatment of *closo*-1,2-$C_2B_{10}H_{12}$ with hot aqueous sodium hydroxide abstracts one boron vertex from the cluster and yields the monoanion [*nido*-7,8-$C_2B_9H_{12}$]⁻. The corresponding dianion is isolated as the sodium salt [*nido*-7,8-$C_2B_9H_{12}$]⁻ reacts with NaH in THF solution. The *nido*-anion [7,8-$C_2B_9H_{11}$]²⁻ (Fig. 5.3) is a particularly useful precursor for the synthesis of heterocarbaboranes and also for boron-functionalized *closo*-1,2-C_2-3-R-$B_{10}H_{11}$. An example of the latter is the reaction of $Na_2[7,8-C_2B_9H_{11}]$ with $PhBCl_2$ giving 1,2-C_2-3-Ph-3-$B_{10}H_{11}$. The pentagonal open-face of the *nido*-cage of [7,8-$C_2B_9H_{11}$]²⁻ resembles a cyclopentadienyl (η^5-Cp) ligand in its ability to bind to e.g. transition metals (see Section 6.2).

One method of synthesizing monocarbaboranes uses cyano-derivatives of boranes. Protonation of *arachno*-[$B_{10}H_{13}(CN)$]²⁻ (in which the cyano-group is terminally bound) leads to a mixture of the carbaboranes *nido*-$C(NH_3)B_{10}H_{12}$ and *nido*-$C(NH_3)B_9H_{11}$ (Fig. 5.4). At 70°C, *nido*-$C(NH_3)B_9H_{11}$ reacts with piperidine, eliminating NH_3 and forming [pipH][*closo*-1-CB_9H_{10}]. Reduction of *nido*-$C(NH_3)B_9H_{11}$ with sodium in liquid ammonia yields, after hydrolysis, *nido*-[CB_9H_{12}]⁻ which is related structurally to its *nido*-precursor (Fig. 5.4) and is an important starting material in monocarbaborane chemistry.

Fig. 5.3 [*nido*-7,8-$C_2B_9H_{11}$]²⁻; the cage atom numbering begins at the apical boron atom opposite to the open face. Note the difference in numbering when the cage is a closed icosahedron (Fig. 3.6).

piperidine (pip)

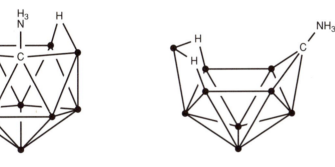

Fig. 5.4 *nido*-$C(NH_3)B_{10}H_{12}$ and *nido*-$C(NH_3)B_9H_{11}$.

Coupling small carbaborane cages together may be achieved by using a platinum(II) bromide catalyst in a manner analogous to the reactions shown in Eqns 5.21–5.23. The catalysed couplings tend to be specific; e.g. coupling of two 1,6-$C_2B_4H_6$ clusters preferentially yields 2,2'-{1,6-$C_2B_4H_5$} with elimination of H_2, and 2,2'-{1,5-$C_2B_3H_4$} is the specific product of the $PtBr_2$ catalysed coupling of two molecules of 1,5-$C_2B_3H_5$. Each coupling is via an *exo*-B–B bond.

5.3 *p*-Block heteroboranes

In this section, selected methods of incorporating *p*-block elements from groups 13 to 16, excluding carbon, into borane clusters are described. Examples have been chosen so as to demonstrate the use of general methodologies.

Aluminium, gallium, and indium

The reaction of diborane(6) with $Al(BH_4)_3$ (Fig. 5.5) or with $AlMe_3$ yields an analogue of B_5H_{11} (Eqns 5.26 and 5.27) with the aluminium atom occupying

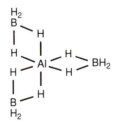

Fig. 5.5 Structure of $Al(BH_4)_3$.

an apical site (site 1) in the *arachno*-framework. The insertion of an {AlH₃}-fragment into pentaborane(9) gives an analogue of B_6H_{12} (Eqn 5.28). The syntheses of analogues of B_4H_{10} with the heteroatom in the wingtip site (site 2, see Fig. 3.8) of the *arachno*-framework are shown in Eqns 5.29–5.31. Me_2MCl (M = Al or Ga) provides a source of $\{Me_2M\}^+$ that adds to $[B_3H_8]^-$ to give *arachno*-2,2-Me₂-2-MB₃H₈.

$$B_2H_6 + 2\,Al(BH_4)_3 \xrightarrow{\text{benzene, 100°C}} 2\,\textit{arachno}\text{-1-AlB}_4H_{11} + 4\,H_2 \qquad \text{Eqn 5.26}$$

$$5\,B_2H_6 + 2\,AlMe_3 \xrightarrow{\text{benzene, 100°C}} 2\,\textit{arachno}\text{-1-AlB}_4H_{11} + 4\,H_2 + 2\,Me_3B \qquad \text{Eqn 5.27}$$

$$B_5H_9 + 2\,AlMe_3 \xrightarrow{\text{benzene, 80°C}} \textit{arachno}\text{-AlB}_5H_{12} + H_2 + B_2H_6 + \text{decomposition products} \qquad \text{Eqn 5.28}$$

$$Ga_2Cl_2H_4 + 2\,[B_3H_8]^- \xrightarrow{-30°C} 2\,\textit{arachno}\text{-2-GaB}_3H_{10} + 2\,Cl^- \qquad \text{Eqn 5.29}$$

$$Me_2AlCl + [B_3H_8]^- \longrightarrow \textit{arachno}\text{-2,2-Me}_2\text{-2-AlB}_3H_8 + Cl^- \qquad \text{Eqn 5.30}$$

$$Me_2GaCl + [B_3H_8]^- \longrightarrow \textit{arachno}\text{-2,2-Me}_2\text{-2-GaB}_3H_8 + Cl^- \qquad \text{Eqn 5.31}$$

The elimination of methane is the driving force for the reaction of $GaMe_3$ or $InMe_3$ with *nido*-C₂B₄H₈ to give *closo*-1-Me-M-2,3-C₂B₄H₆ (M = Ga or In; Fig. 3.7). Similarly, loss of ethane drives the reaction of $AlEt_3$ or $GaEt_3$ with *arachno*-7,8-C₂B₉H₁₃ to give *nido*-7,8-C₂B₉H₁₂(MEt₂) which, on heating, eliminates a second mole of ethane to yield *closo*-1-Et-1-M-2,3-C₂B₉H₁₁ (M = Al or Ga).

Silicon, tin, germanium, and lead

The first silicon analogue of an icosahedral dicarbaborane was prepared in 1990 from $B_{10}H_{14}$ (Eqn 5.32). 1,2-Me₂-1,2-Si₂B₁₀H₁₀ (Fig. 3.7) is stable to both air and moisture but less stable than its carbon counterpart. It is possible to introduce a silicon atom that does not carry a terminal substituent; two environments are exemplified in the products shown in Eqns 5.33 and 5.34. The *nido*-dianionic precursors react with silicon tetrachloride to give either a single *closo*-cluster as in the case of 1-Si-2,3-(SiMe₃)₂-2,3-C₂B₄H₄ or a *commo*-cluster in which the silicon atom is sited at the point of cage coupling (see Fig. 3.9). In Eqn 5.33 where there are two competitive reaction paths, it is the *commo*-product that predominates. Similar reactions occur between *nido*-[2,3-(SiMe₃)₂-2,3-C₂B₄H₅]⁻ and $GeCl_4$ or $SnCl_2$ (Eqns 5.35 and 5.36). When *closo*-1-Sn-2,3-(SiMe₃)₂-2,3-C₂B₄H₄ is treated with $GeCl_4$ at 150°C, the tin atom is replaced by a germanium atom but the substitution is accompanied by a change from a *closo*- to *commo*-derivative, 2,2',3,3'-(SiMe₃)₄-*commo*-1,1'-Ge(1-Ge-2,3-C₂B₄H₄)₂ (Fig. 5.6).

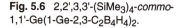

Fig. 5.6 2,2',3,3'-(SiMe₃)₄-*commo*-1,1'-Ge(1-Ge-2,3-C₂B₄H₄)₂.

$$2\,B_{10}H_{14} + 2\,MeHSi(NMe_2)_2 \xrightarrow[\text{benzene}]{\text{Reflux in}} 1\text{,2-Me}_2\text{-Si}_2B_{10}H_{10} + 6\text{,9-(Me}_2\text{NH)}_2B_{10}H_{12} + 2\,H_2 + 2\,Me_2NH \qquad \text{Eqn 5.32}$$

$$NaLi[2,3\text{-R}_2C_2B_4H_4] + SiCl_4 \xrightarrow[-\,LiCl\ -\,NaCl]{\text{0°C THF}} \textit{closo}\text{-1-Si-2,3-R}_2\text{-2,3-C}_2B_4H_4 \qquad \text{Eqn 5.33}$$
$$R = SiMe_3 \qquad\qquad\qquad\qquad + 2,2',3,3'\text{-R}_4\text{-}\textit{commo}\text{-1,1'-Si(1-Si-2,3-C}_2B_4H_4)_2$$

$$\text{Li}_2[\text{C}_2\text{B}_9\text{H}_{11}] + \text{SiCl}_4 \xrightarrow[\text{– LiCl}]{\text{Reflux in benzene}} \textit{commo-} 3,3'\text{-Si}(3\text{-Si-}1,2\text{-C}_2\text{B}_9\text{H}_{11})_2 \qquad \textbf{Eqn 5.34}$$

$$\text{Na}[2,3\text{-R}_2\text{C}_2\text{B}_4\text{H}_5] + \text{GeCl}_4 \xrightarrow[\text{– NaCl}]{\text{0°C THF}} \textit{closo-}1\text{-Ge-}2,3\text{-R}_2\text{-}2,3\text{-C}_2\text{B}_4\text{H}_4 \qquad \textbf{Eqn 5.35}$$
$$\text{R} = \text{SiMe}_3 \qquad\qquad + 2,2',3,3'\text{-R}_4\text{-}\textit{commo-}1,1'\text{-Ge}(1\text{-Ge-}2,3\text{-C}_2\text{B}_4\text{H}_4)_2$$

$$\text{Na}[2,3\text{-R}_2\text{C}_2\text{B}_4\text{H}_5] + \text{SnCl}_2 \xrightarrow[\text{– NaCl}]{\text{0°C THF}} \textit{closo-}1,1\text{-(THF)}_2\text{-1-Sn-}2,3\text{-R}_2\text{-}2,3\text{-C}_2\text{B}_4\text{H}_4 \qquad \textbf{Eqn 5.36}$$
$$\text{R} = \text{SiMe}_3$$

The introduction of lead is considered in Fig. 6.12.

Nitrogen

Despite the large difference in electronegativity between boron and nitrogen, it is possible to introduce nitrogen as a heteroatom into a borane cluster without unduly perturbing the electronic structure of the cluster. One potential source of nitrogen in the synthesis of azaboranes is the nitrite ion. The reaction of $\text{B}_{10}\text{H}_{14}$ with sodium nitrite in THF gives an intermediate compound proposed as $\text{Na}[\text{B}_{10}\text{H}_{12}\text{NO}_2]$. Protonation of the intermediate occurs in concentrated H_2SO_4 to yield *nido*-6-(NH)B_9H_{11} (structurally related to $\text{B}_{10}\text{H}_{14}$) and in dilute aqueous HCl to give *arachno*-4-(NH)B_8H_{12}.

An {NH}-unit is isolobal with a {BH^{2-}}-unit and so the introduction of one nitrogen vertex into a *closo*-hydroborate dianion will yield a neutral azaborane. The reaction sequence in Eqn 5.37 illustrates the use of the azide ion as a source of a cluster nitrogen atom. The nitrogen atom in the icosahedral *closo*-$\text{B}_{11}\text{H}_{11}\text{NH}$ is in an unusual environment with a connectivity of six; it interacts with one terminal hydrogen atom and five cage boron atoms.

> Pauling electronegativities of B and N are 2.0 and 3.0, respectively.

> $\text{B}_{10}\text{H}_{12}(\text{SMe}_2)_2$ (Eqn 5.37) may be prepared from $\text{B}_{10}\text{H}_{14}$ by reaction with SMe$_2$; see Eqn 5.25.

$$\textit{arachno-}\text{B}_{10}\text{H}_{12}(\text{SMe}_2)_2 \xrightarrow[\text{– 2 SMe}_2]{\text{HN}_3} \textit{arachno-}\text{B}_{10}\text{H}_{12}(\text{N}_3)(\mu\text{-NH}_2) \xrightarrow{\Delta} \textit{nido-}\text{B}_{10}\text{H}_{12}\text{NH} \qquad \textbf{Eqn 5.37}$$
$$\downarrow \text{Et}_3\text{N.BH}_3$$
$$\textit{closo-}\text{B}_{11}\text{H}_{11}\text{NH} \xleftarrow{\text{H[BF}_4]} [\text{Et}_3\text{NH}][\text{B}_{11}\text{H}_{11}\text{N}]$$

In Chapter 4, the general guidelines for *PSEPT* included the statement that, with the exception of capping vertices, it is usual to remove the vertex of highest connectivity in going from a *closo*- to *nido*-cluster. The azaborane *nido*-$\text{B}_4\text{Me}_4\text{N}_2{}^t\text{Bu}_2$ is an exception (Fig. 5.7); it exhibits a vacant equatorial rather than apical site in the parent pentagonal bipyramidal skeleton.

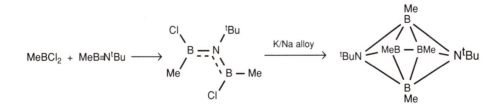

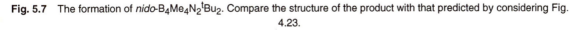

Fig. 5.7 The formation of *nido*-B$_4$Me$_4$N$_2{}^t$Bu$_2$. Compare the structure of the product with that predicted by considering Fig. 4.23.

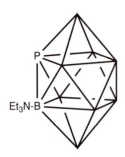

Fig. 5.8 *closo*-6-Et$_3$N-2-PB$_9$H$_8$; each unmarked vertex = BH.

Phosphorus

The number of known binary phosphaboranes (i.e. cages consisting of only P and B atoms) is small. A suitable source of phosphorus is PCl$_3$. The reaction of PCl$_3$ with B$_{10}$H$_{14}$ (Eqn 5.38) gives *closo*-1,2-P$_2$B$_{10}$H$_{10}$. This route is analogous to those used for the introduction of the heavier group 15 elements (Eqns 5.41–5.43). A second *closo*-phosphaborane (Fig. 5.8) is also formed in the reaction. Pyrolysis of *closo*-1,2-P$_2$B$_{10}$H$_{10}$ at ≈590°C causes isomerization to *closo*-1,7-P$_2$B$_{10}$H$_{10}$. The icosahedral cluster is a precursor to smaller cages (e.g. Eqn 5.39). In each of the phosphaborane clusters shown, the phosphorus atom carries one pair of electrons and no terminal substituent. Use of an alkyl or aryl phosphorus dichloride permits the introduction of an alkyl or aryl substituted phosphorus atom as indicated in Eqn 5.40.

$$B_{10}H_{14} \xrightarrow[\substack{\text{in the presence} \\ \text{of Et}_3\text{N and Na[BH}_4]}]{\text{PCl}_3 \quad \text{THF}} \text{\textit{closo}-1,2-P}_2\text{B}_{10}\text{H}_{10} + \text{\textit{closo}-6-Et}_3\text{N-2-PB}_9\text{H}_8 \qquad \textbf{Eqn 5.38}$$

$$\text{\textit{closo}-1,2-P}_2\text{B}_{10}\text{H}_{10} \xrightarrow[\text{2. HCl}]{\text{1. NaOH (aq)}} [\text{\textit{nido}-7-PB}_{10}\text{H}_{12}]^- \qquad \textbf{Eqn 5.39}$$

$$B_{10}H_{14} \xrightarrow[\text{R = Me, Et, Pr, Ph}]{\text{RPCl}_2; \text{ excess NaH}} [\text{\textit{nido}-7-R-7-PB}_{10}\text{H}_{11}]^- \xrightarrow{\text{H}^+} \text{\textit{nido}-7-R-7-PB}_{10}\text{H}_{12} \qquad \textbf{Eqn 5.40}$$

Arsenic, antimony, and bismuth

The halides of the later group 15 elements are useful precursors to heteroboranes incorporating arsenic, antimony, or bismuth. In the presence of base, *nido*-B$_{10}$H$_{14}$ reacts with EX$_3$ (E = As, Sb, or Bi; X = halogen) to yield *closo*-1,2-E$_2$B$_{10}$H$_{10}$ (Eqns 5.41–5.43). The reaction shown in Eqn 5.42 is complicated by the formation of [*nido*-7-AsB$_{10}$H$_{12}$]$^-$. This can be specifically prepared from B$_{10}$H$_{14}$ and AsI$_3$ in the presence of base and a reducing agent such as borohydride ion. The ion [*nido*-7-AsB$_{10}$H$_{12}$]$^-$ reacts further with AsI$_3$ in the presence of Et$_3$N to give *closo*-1,2-As$_2$B$_{10}$H$_{10}$.

$$B_{10}H_{14} \xrightarrow[\text{in the presence of Et}_3\text{N}]{\text{BiCl}_3, \text{ THF, 25°C}} \text{\textit{closo}-1,2-Bi}_2\text{B}_{10}\text{H}_{10} \qquad \textbf{Eqn 5.41}$$

$$B_{10}H_{14} \xrightarrow[\text{in the presence of Et}_3\text{N}]{\text{BiCl}_3, \text{ AsCl}_3, \text{ THF, 25°C}} \text{\textit{closo}-1,2-Bi}_2\text{B}_{10}\text{H}_{10} + \text{\textit{closo}-1-As-2-BiB}_{10}\text{H}_{10}$$
$$+ \text{\textit{closo}-1,2-As}_2\text{B}_{10}\text{H}_{10} + [\text{\textit{nido}-7-AsB}_{10}\text{H}_{12}]^- \qquad \textbf{Eqn 5.42}$$

$$B_{10}H_{14} \xrightarrow[\text{in the presence of Et}_3\text{N}]{\text{BiCl}_3, \text{ SbI}_3, \text{ THF, 25°C}} \text{\textit{closo}-1,2-Bi}_2\text{B}_{10}\text{H}_{10} + \text{\textit{closo}-1-Bi-2-SbB}_{10}\text{H}_{10} + \text{\textit{closo}-1,2-Sb}_2\text{B}_{10}\text{H}_{10} \qquad \textbf{Eqn 5.43}$$

Sulfur, selenium, and tellurium

Polysulfides such as K$_2$S$_n$ are used as a source of cluster sulfur. Ammonium sulfide is also used; this decomposes to give NH$_3$, [NH$_4$]SH$_n$ and polysulfides, [S$_n$]$^{2-}$. In Eqn 5.44, the polysulfide is represented simply as S^{2-}.

One source of a cluster-bound sulfur atom is a polysulfide. Reaction of decaborane(14) with ammonium sulfide in the presence of water results in the concomitant abstraction of one boron vertex and incorporation of a sulfur atom (Eqn 5.44). The resulting *arachno*-cluster anion [B$_9$H$_{12}$S]$^-$ is an important precursor to other thiaboranes, and examples are given in Eqns 5.45 and 5.46.

Oxyanions of the group 16 elements are also used to synthesize heteroborane clusters (Eqns 5.47–5.49). As in the previous examples, the decaborane cage is partially degraded as the heteroatom is introduced. In the reaction between $B_{10}H_{14}$ and $[S_2O_5]^{2-}$, the degree of cluster degradation depends upon the acidic medium present. Degradation does not always occur; in the reaction of $[B_{11}H_{14}]^-$ with $NaHSeO_3$, *closo*-1-$SeB_{11}H_{11}$ forms and with TeO_2, *closo*-1-$TeB_{11}H_{11}$ is produced. Use of the polyselenide and polytelluride cations $[Se_4]^{2+}$ and $[Te_4]^{2+}$ is shown in Eqns 5.50 and 5.51.

In each of the thia-, selena-, and telluraboranes shown in Eqns 5.44–5.51, the group 16 atom is devoid of a terminal substituent but carries a lone pair of electrons.

$$B_{10}H_{14} + S^{2-} + 4\,H_2O \longrightarrow [\textit{arachno}\text{-}6\text{-}SB_9H_{12}]^- + [B(OH)_4]^- + 3\,H_2 \qquad \textbf{Eqn 5.44}$$

$$2\,[\textit{arachno}\text{-}6\text{-}SB_9H_{12}]^- + I_2 \xrightarrow{\text{Reflux in benzene}} \textit{nido}\text{-}6\text{-}SB_9H_{11} \xrightarrow[\text{on cold finger at }-78°C]{450°C;\text{ product collected}} \textit{closo}\text{-}1\text{-}SB_9H_9 \qquad \textbf{Eqn 5.45}$$

$$2\,[\textit{arachno}\text{-}6\text{-}SB_9H_{12}]^- \xrightarrow{200°C} [\textit{nido}\text{-}7\text{-}SB_{10}H_{11}]^- \xrightarrow{H^+} \textit{nido}\text{-}7\text{-}SB_{10}H_{12} \xrightarrow[\substack{\text{collected on}\\\text{cold finger at }-78°C}]{380°C;\text{ product}} \textit{closo}\text{-}1\text{-}SB_{11}H_{11} \qquad \textbf{Eqn 5.46}$$

$$B_{10}H_{14} + K_2S_2O_5\,(aq) \xrightarrow{\text{Conc. }H_2SO_4} \textit{nido}\text{-}6\text{-}SB_9H_{11} \qquad \textbf{Eqn 5.47}$$

$$B_{10}H_{14} + K_2S_2O_5\,(aq) \xrightarrow{\text{HCl(aq)}} \textit{arachno}\text{-}4\text{-}SB_8H_{12} \qquad \textbf{Eqn 5.48}$$

$$B_{10}H_{14} + Na_2Se_2O_3 \xrightarrow{\text{THF, HCl(aq)}} \textit{nido}\text{-}7,8\text{-}Se_2B_9H_9 \qquad \textbf{Eqn 5.49}$$

$$B_{10}H_{14} + Na_2Se_4 \longrightarrow \textit{nido}\text{-}7,8\text{-}Se_2B_9H_9 + [\textit{nido}\text{-}7\text{-}SeB_{10}H_{11}]^- \xrightarrow{H^+} \textit{nido}\text{-}7\text{-}SeB_{10}H_{12} \qquad \textbf{Eqn 5.50}$$

$$B_{10}H_{14} + Na_2Te_4 \longrightarrow [\textit{nido}\text{-}7\text{-}TeB_{10}H_{11}]^- \xrightarrow{H^+} \textit{nido}\text{-}7\text{-}TeB_{10}H_{12} \qquad \textbf{Eqn 5.51}$$

5.4 Clusters of boron atoms other than the boranes

Boron halide and related clusters

Most neutral B_nX_n clusters are synthesized from B_2X_4 by thermally initiated disproportionation into BX_3 and B_nX_n; for X = Cl, n = 8–12, for X = Br, n = 7–10, and for X = I, n = 8 and 9. Coupled clusters are also present in the product mixture. The proportions of products in a given reaction mixture can be altered by varying the reaction conditions. For example, heating B_2Cl_4 at 450°C for a few minutes gives B_9Cl_9 in high yield, whereas heating B_2Cl_4 at 100°C for several days in the presence of CCl_4 provides a high yield route to B_8Cl_8. Mixed substituents may be introduced (Eqn 5.52).

The tetrahedral cluster B_4Cl_4 is not formed from B_2Cl_4 but is instead prepared by an electrical discharge through BCl_3 in the presence of mercury. Yields are small, but are improved by using a radio-frequency discharge. B_4Cl_4 has been used to access $B_4{}^tBu_4$ (Eqn 5.53); a more recent synthesis is shown in Eqn 5.54.

$$B_9Br_9 \xrightarrow{TiCl_4} B_9Br_{9-n}Cl_n \qquad \textbf{Eqn 5.52}$$

$$B_4Cl_4 + 4\,{}^tBuLi \longrightarrow B_4{}^tBu_4 + 4\,LiCl \qquad \textbf{Eqn 5.53}$$

$$12\,{}^tBuBF_2 + 8\,M \longrightarrow B_4{}^tBu_4 + 8\,M[\,{}^tBuBF_3] \qquad \textbf{Eqn 5.54}$$
$$\text{M = Na/K alloy}$$

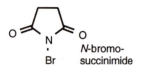

N-bromo-
succinimide

Anionic boron halide clusters are prepared by the halogenation of the corresponding *closo*-hydroborate dianion (Eqns 5.55–5.58). The chloroderivative $[B_9Cl_9]^{2-}$ may also be synthesized by using *N*-chlorosuccinimide in a reaction analogous to that given for the bromo-cluster in Eqn 5.57.

$$[B_6H_6]^{2-} \xrightarrow{\quad X_2 (X = Cl, Br, I), \ aq. \ alkali \quad} [B_6X_6]^{2-} \quad \textbf{Eqn 5.55}$$

$$[B_9H_9]^{2-} \xrightarrow{\quad SO_2Cl_2 \quad} [B_9Cl_9]^{2-} \quad \textbf{Eqn 5.56}$$

$$[B_9H_9]^{2-} \xrightarrow{\quad N\text{-bromosuccinimide} \quad} [B_9Br_9]^{2-} \quad \textbf{Eqn 5.57}$$

$$[B_9H_9]^{2-} \xrightarrow{\quad Excess \ I_2 \quad} [B_9I_9]^{2-} \quad \textbf{Eqn 5.58}$$

Clusters with boron–nitrogen or boron–phosphorus bonds

Monocyclic borazines $[RBNR']_3$ (R and R' = alkyl or aryl) are typically prepared by reacting a primary amine $R'NH_2$ with an organoborane, RBH_2. The elimination of H_2 is the driving force for the process. Oligomerization to form cluster compounds does not occur. Similarly, the reaction of BCl_3 with NCl_3 leads only to the trimer $[ClBNCl]_3$. In the case of the hydrazine adduct $H_4N_2.B^tBuH_2$, dimerization accompanied by H_2 elimination leads to a six-membered ring but further oligomerization occurs to generate a cluster (Fig. 5.9). The cubane-like tetramer $[MeNBCl_2]_4$ (Fig. 3.13) is formed by the dimerization of the open chain compound $Cl_2BN(Me)N(Me)BCl_2$. A closely related boron–phosphorus cluster is formed according to Fig. 5.10.

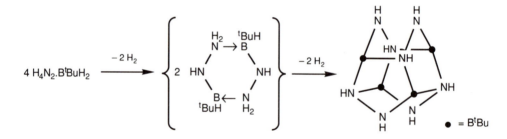

Fig. 5.9 The formation of the tetrameric cluster $[^tBuBN_2H_2]_4$.

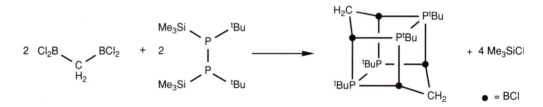

Fig. 5.10 The formation of the cubane-like cluster $[^tBuPB(Cl)CH_2B(Cl)P^tBu]_2$.

5.5 Aluminium

The icosahedral Al_{12}-cluster core

It seems surprising that, despite the fact that an aluminium atom can be introduced into several borane clusters without perturbation of the cluster structure, clusters with all-aluminium cores are represented only by

$[Al_{12}{}^iBu_{12}]^{2-}$. The synthesis of the $[Al_{12}{}^iBu_{12}]^{2-}$ dianion is given in Eqn 5.59. The cluster anion is quite stable in air and under an inert atmosphere it is thermally stable up to 150°C.

$$12\ {}^iBu_2AlCl \xrightarrow[\text{O°C (3 days), 25°C (1 day)}]{\text{K in hexane}} K_2[\,Al_{12}{}^iBu_{12}] \quad \text{Low yield} \qquad \textbf{Eqn 5.59}$$

Iminoalanes

The synthetic strategies used to form iminoalanes (Figs. 3.14 and 3.15) centre on Lewis acid–base interactions between aluminium and nitrogen species. Selected general routes are given in Eqns 5.60–5.63. The degree of aggregation in the cluster depends on the alkyl or aryl substituents and on reaction conditions. In Eqn 5.60, the stability of the intermediate compound is determined by the identity of R, e.g. if R = H, dihydrogen elimination is facile and the intermediate is not stable.

$$n\,AlR_3\ +\ n\,R'NH_2 \xrightarrow[-\,RH]{} \{[R_2AlNHR']_n\} \xrightarrow[-\,RH]{} [RAlNR']_n \quad \text{R and R' = alkyl or aryl} \qquad \textbf{Eqn 5.60}$$

$$n\,M[AlH_4]\ +\ n\,RNH_2 \longrightarrow [HAlNR]_n\ +\ n\,MH\ +\ 2n\,H_2 \qquad \text{M = Li, Na; R = } {}^iPr,\ {}^nBu,\ {}^tBu \qquad \textbf{Eqn 5.61}$$

$$n\,Li[AlH_4]\ +\ n\,RNH_3Cl \longrightarrow [HAlNR]_n\ +\ n\,LiCl\ +\ 3n\,H_2 \qquad \text{R = various alkyl} \qquad \textbf{Eqn 5.62}$$

$$2n\,Al\ +\ 2n\,RNH_2 \xrightarrow{\text{High temp; high H}_2\ \text{pressure}} 2\,[HAlNR]_n\ +\ n\,H_2 \qquad \text{R = various alkyl} \qquad \textbf{Eqn 5.63}$$

The pyrolysis of the adduct $Me_3Al.NH_2Me$ at 215°C leads to a mixture of $[MeAlNMe]_8$ and $[MeAlNMe]_7$ (Fig. 3.14). At the lower temperature of 175°C, $(MeAlNMe)_6(Me_2AlNHMe)_2$ (Fig. 3.15) becomes the dominant product. This cluster has been shown to be an intermediate in the formation of the heptamer and octamer; after isolation of $(MeAlNMe)_6(Me_2AlNHMe)_2$, a sample heated to 215°C converts to $[MeAlNMe]_7$ and to small amounts of $[MeAlNMe]_8$. The reaction can be monitored by ^1H NMR spectroscopy and the appearance of the heptamer approximately mirrors the decay of $(MeAlNMe)_6(Me_2AlNHMe)_2$.

The iminoalane $(HAlN^iPr)_2(H_2AlNH^iPr)_3$ (Fig 3.15) is synthesized by reacting aluminium hydride with ^{i}propylamine in diethyl ether under reflux. The product has a more open cage-structure than many other iminoalanes.

An aluminaphosphacubane

Despite the range of iminoalane cluster molecules, phosphorus analogues have not been forthcoming until recently. The reaction in Eqn 5.64 provides a route to the aluminaphosphacubane $[{}^iBuAlP(SiPh_3)]_4$. The tetramer is subject to both electrophilic and nucleophilic attack since the Al–P bonds are polarized and the cubane is degraded by ethanol to give $^iBuAl(OEt)_2$ and Ph_3SiPH_2. This point emphasizes how critical the choice of solvent for a synthesis can be.

Pauling electronegativity values for Al and P are 1.61 and 2.19, respectively.

$$4\,{}^iBu_2AlH\ +\ 4\,(Ph_3Si)PH_2 \xrightarrow[-\,H_2]{25\text{°C, toluene}} 4\,{}^iBu_2AlPH(SiPh_3) \xrightarrow[-\,{}^iBuH]{\substack{\text{Reflux in}\\\text{toluene}}} [{}^iBuAlP(SiPh_3)]_4 \qquad \textbf{Eqn 5.64}$$

5.6 Gallium and indium

The iminogallane $(MeGaNMe)_6(Me_2GaNHMe)_2$ is prepared in a similar way to the analogous iminoalane (see above) by heating a mixture of Me_3Ga and $MeNH_2$ at 210°C.

The adamantane-like sulfide and selenide octa-anions, $[E_4X_{10}]^{8-}$ (E = Ga or In, X = S or Se) are prepared from simple sulfides or selenides of the appropriate group 13 element; Eqn 5.65 shows the formation of $[In_4Se_{10}]^{8-}$.

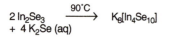

$$2\,In_2Se_3 + 4\,K_2Se\,(aq) \xrightarrow{90°C} K_8[In_4Se_{10}]$$

Eqn 5.65

5.7 Thallium

Thallium(I) alkoxides

The aerial oxidized alcoholysis of metallic thallium provides a synthetic pathway to the tetrameric thallium(I) ethoxide (Eqn 5.66). Other derivatives can be prepared by the replacement of the alkoxide substituent, e.g. $[TlOMe]_4$ is obtained by treating $[TlOEt]_4$ with methanol.

$$4\,Tl + 4\,EtOH \rightarrow [TlOEt]_4 + 2\,H_2$$

Eqn 5.66

Thallium(I) thiolates

Until 1989, structurally characterized thallium(I) thiolates were not known. The reaction of a thallium(I) salt with an alkali metal thiolate, M[SR], (Eqn 5.67) offers a route to a compound, the empirical formula of which implies a simple thallium(I) thiolate TlSR. As in thallium(I) alkoxides, the thiolates are molecular clusters, the structures of which depend on R. For R = ^tBu, a double-cubane (Fig. 3.18) is formed.

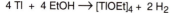

$$Tl_2CO_3 + NaSR \longrightarrow \begin{array}{l} 2\,TlSR \\ + Na_2CO_3 \\ (R = Ph,\,^tBu) \end{array}$$

Eqn 5.67

Cubanes with thallium–nitrogen bonds

The cubane clusters $Tl_2(MeSi)_2(N^tBu)_4$ and $Tl_6(MeSi)_2(N^tBu)_6$ (Fig. 3.17) are synthesized by thallium-for-lithium exchange (Eqns 5.68 and 5.69). The lithium-containing precursors are structurally related to the thallium products and are formed by the lithiation of an appropriate silylamine (Eqn 5.70). Association of the lithiated silylamines generates silazane clusters that function as templates in the syntheses of related compounds. The origin of the double cubane-type of structure for both $Li_6(MeSi)_2(N^tBu)_6$ and $Tl_6(MeSi)_2(N^tBu)_6$ can be appreciated by studying Fig. 5.11.

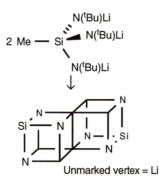

Unmarked vertex = Li

Fig. 5.11 Derivation of the core of $Li_6(MeSi)_2(N^tBu)_6$ by dimerization of the lithiated silylamine precursor. In reality the cluster core is not a perfect double-cube.

$$Li_2(MeSi)_2(N^tBu)_4 + 2\,TlCl \longrightarrow Tl_2(MeSi)_2(N^tBu)_4 + 2\,LiCl \qquad \textbf{Eqn 5.68}$$

$$Li_6(MeSi)_2(N^tBu)_6 + 6\,TlCl \longrightarrow Tl_6(MeSi)_2(N^tBu)_6 + 6\,LiCl \qquad \textbf{Eqn 5.69}$$

$$2\,MeSi(N^tBuH)_3 + 6\,^nBuLi \longrightarrow [MeSi(N^tBuLi)_3]_2 \qquad \textbf{Eqn 5.70}$$
$$(\equiv Li_6(MeSi)_2(N^tBu)_6) + 6\,^nBuH$$

5.8 Carbon clusters

Tetra^{t}butyltetrahedrane

The successful synthesis of a tetrahedrane cluster is critically dependent on the *exo*-substituent. Tetra^{t}butyltetrahedrane is formed as a colourless, air stable crystalline compound after the photolysis of 2,3,4,5-tetra^{t}butylcyclo-penta-2,4-dienone (Fig. 5.12), although the photolysis is not as simple as Fig 5.12 might suggest. The tetrahedrane cluster converts to tetra^{t}butyl-cyclobutadiene on heating at 130°C; the process is reversed upon photolysis.

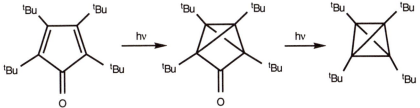

Fig. 5.12 Synthesis of $C_4{}^t Bu_4$.

Benzvalene and prismane—isomers of benzene

Benzvalene, characterized by a foul odour, is formed by photolysing benzene but since the photochemical decomposition of benzvalene is sensitized by benzene, this is not a good route. A better method is shown in Eqn 5.71.

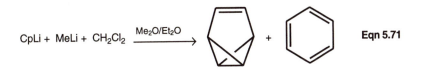

$$CpLi + MeLi + CH_2Cl_2 \xrightarrow{Me_2O/Et_2O}$$

Eqn 5.71

Prismane is an explosive liquid; it is stable at room temperature but isomerizes to benzene at 90°C. The reaction of benzvalene with the dienophile shown in Fig. 5.13 begins the closure of the C_6-skeleton. The loss of dinitrogen is the driving force for the final stage of the synthesis.

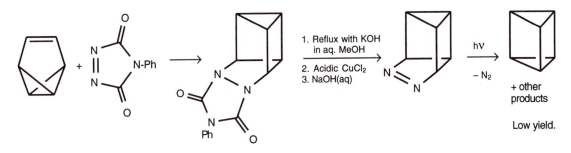

Fig. 5.13 A synthesis of prismane. Both the precursor and product are isomers of benzene.

Cubane

Cubane, C_8H_8, may be formed from a norbornene precursor by a multistep procedure that allows the gradual closure of the cage. A more convenient synthesis begins with half of the cubane in the form of the cyclobutadiene ligand in the organometallic complex $(\eta^4\text{-}C_4H_4)Fe(CO)_3$. A Diels–Alder addition of 2,5-dibromo-*p*-benzoquinone is the start of the closure of the C_8-cage (Fig. 5.14).

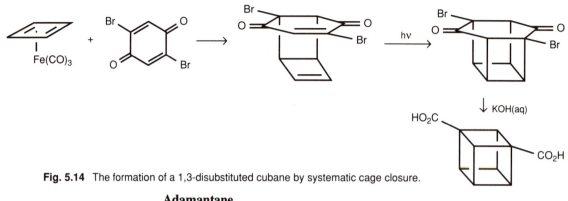

Fig. 5.14 The formation of a 1,3-disubstituted cubane by systematic cage closure.

Adamantane

There are various methods of preparing adamantane and its derivatives. One general approach is that of ring closure via the insertion of a methylene group into a bicyclo[3.3.1] precursor. A better strategy is that of isomerization which relies upon the fact that the adamantane cage is a particularly favourable structure and an appropriate precursor will undergo rearrangement to form the adamantane cage (Eqn 5.72).

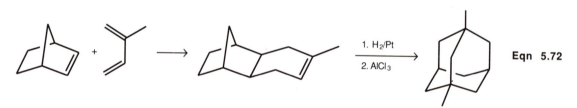

$$\text{Eqn 5.72}$$

5.9 Silicon

Simple silicon cages

A metal or glass is annealed *when it is heated and cooled in a controlled way to relieve stresses.*

When silicon reacts with a mixture of lithium and potassium at 800°C followed by annealing and slow cooling of the products, red alkali metal silicides $K_3Li[Si_4]$ and $K_7Li[Si_4]_2$ are isolated. These compounds are inflammable in air.

$$4\ Ph_2(^tBuMe_2Si)Si\text{–}Si(SiMe_2{}^tBu)Ph_2 \xrightarrow[+\ 16\ HBr]{\begin{array}{c}AlBr_3\\ -\ 16\ C_6H_6\end{array}} 4\ Br_2(^tBuMe_2Si)Si\text{–}Si(SiMe_2{}^tBu)Br_2 \xrightarrow[-\ 16\ NaBr]{Na} Si_8(SiMe_2{}^tBu)_8 \quad \textbf{Eqn 5.73}$$

$$8\ (^tBuMe_2Si)SiPh_3 + 8\ HBr \xrightarrow[-\ 8\ C_6H_6]{AlBr_3} 8\ (^tBuMe_2Si)SiBr_3 \xrightarrow[-\ 24\ NaBr]{Na} Si_8(SiMe_2{}^tBu)_8 \quad \textbf{Eqn 5.74}$$

Polycyclic silanes such as Si_7Me_{12}, Si_8Me_{14}, $Si_{10}Me_{16}$, $Si_{11}Me_{18}$, and $Si_{13}Me_{22}$ are formed by treating mixtures of $MeSiCl_3$ and Me_2SiCl_2 with sodium/potassium alloy in the presence of naphthalene. The octasilacubane $Si_8(Si^tBuMe_2)_8$ may be synthesized either by the aggregation of di- or mono-silicon units (Eqns 5.73 and 5.74). The cubane forms thermochromic crystals that are bright yellow at room temperature and colourless at −196°C. At 280°C, an orange glass is formed.

Polycyclic siloxanes and silazanes

Polycyclic siloxanes or silsesquioxanes of the general type $[RSiO_{1.5}]_n$ are synthesized from organochlorosilanes (Eqn 5.75). The cohydrolysis of two differently substituted organochlorosilanes gives siloxanes with mixed substituents. For alkoxysilanes one successful route is the reaction of $(ClSi)_8O_{12}$ with MeONO during which NOCl is eliminated and the terminal chlorides are replaced by methoxy-groups. The choice of precursor for the introduction of the methoxy-groups is critical since ionic reagents (e.g. MeOLi) degrade the siloxane cluster core.

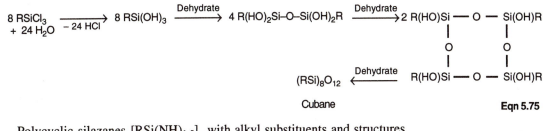

Eqn 5.75

Cubane

Polycyclic silazanes $[RSi(NH)_{1.5}]_n$ with alkyl substituents and structures related to the siloxanes described above are prepared by the general method given in Eqn 5.76. The cubane ($n = 8$) is favoured for R = ^{n}octyl whereas hexagonal prismatic clusters ($n = 6$) are observed for R = methyl, ethyl, ^{n}heptyl, or ^{n}nonyl.

See Section 5.11 for the syntheses of polycyclic silaphosphanes.

$$2n\,RSiCl_3 + 9n\,NH_3 \longrightarrow 2\,[RSi(NH)_{1.5}]_n + 6n\,NH_4Cl$$

R = alkyl from methyl to nonyl   **Eqn 5.76**

5.10 Germanium, tin, and lead

Zintl ions

Early methods for the formation of Zintl ions involved the dissolution of metallic germanium, tin, or lead in liquid ammonia containing sodium. This solvent system is a reducing medium; ionization of Na to Na^+ occurs and the liquid ammonia solvates the free electrons. The present synthetic approach is to extract the alkali metal M from a Zintl phase M_nE_x by using 2,2,2-crypt. This macrocyclic ligand forms a cage around the alkali metal ion and strips it from the Zintl phase (Eqn 5.77). The product is solvated with 1,2-diaminoethane—a crucial solvent for the process (Eqns 5.78-5.84). The stoichiometry of the Zintl phase affects the Zintl ion formed; Eqn 5.78 shows a reaction to form $[Sn_5]^{2-}$ but if the tin content of the Zintl phase in increased, $[Sn_9]^{4-}$ will be favoured. Zintl ions are usually highly coloured.

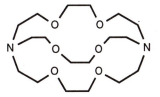

2,2,2-cryptand or 2,2,2-crypt

4,7,13,16-hexaoxa-1,10-diaza-bicyclo[8.8.8]hexacosane

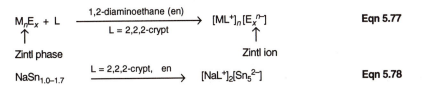

Eqn 5.77

Eqn 5.78

$$NaPb_{1.7-2.0} \xrightarrow{L,\ en} [NaL^+]_2[Pb_5^{2-}] \qquad \textbf{Eqn 5.79}$$

$$KPbSb \xrightarrow{L,\ en} [KL^+]_2[Pb_2Sb_2^{2-}] \qquad \textbf{Eqn 5.80}$$

$$KSn_2 + K_3Bi_2 \xrightarrow{L,\ en} [KL^+]_2[Sn_2Bi_2^{2-}] \qquad \textbf{Eqn 5.81}$$

$$KGe \xrightarrow{L,\ en} [KL^+]_6[Ge_9^{4-}][Ge_9^{2-}] \qquad \textbf{Eqn 5.82}$$

$$NaSnGe \xrightarrow{L,\ en} [NaL^+]_4[Sn_{9-x}Ge_x^{4-}] \qquad \textbf{Eqn 5.83}$$

$$KTlSn \xrightarrow{L,\ en} [KL^+]_3[TlSn_9^{3-}]_{0.5}[TlSn_8^{3-}]_{0.5} \qquad \textbf{Eqn 5.84}$$

L = 2,2,2-crypt

Alkali metal naphthalides:

Naphthalene

$$M + C_{10}H_8 \rightarrow M[C_{10}H_8]$$
$$M = Li,\ Na$$

The electron removed from metal M is delocalized over the aromatic π-system of the naphthalide radical ion. Metal naphthalides are more potent metallating agents than the alkali metal functioning alone.

Simple germanium clusters with *exo*-substituents

In Eqns 5.73 and 5.74, the formation of an octasilacubane cluster was achieved via the aggregation of four dinuclear or eight mononuclear units. Similarly, both mono- and di-germanium fragments can be used as precursors to clusters with Ge_x-cores. A feature of the syntheses of $Ge_6\{CH(SiMe_3)_2\}_6$ and $Ge_8{}^tBu_8Br_2$ is the elimination of lithium halide. The prismane-like $Ge_6\{CH(SiMe_3)_2\}_6$ forms in the reaction of elemental lithium with $Ge\{CH(SiMe_3)_2\}Cl_3$. Reaction of $\{GeBr_2{}^tBu\}_2$ with lithium naphthalide (Eqn 5.85) leads, not to an octagermacubane as may be expected (compare Eqn 5.73), but to an unusual cluster, $Ge_8{}^tBu_8Br_2$, (Fig. 5.15). Both $Ge_6\{CH(SiMe_3)_2\}_6$ and $Ge_8{}^tBu_8Br_2$ are surprisingly stable in air and $Ge_8{}^tBu_8Br_2$ exhibits a high thermal stability.

$$4\{GeBr_2{}^tBu\}_2 + 14\ Li[C_{10}H_8] \rightarrow Ge_8{}^tBu_8Br_2 + 14\ LiBr + 14\ C_{10}H_8 \qquad \textbf{Eqn 5.85}$$

Fig. 5.15 The core of $Ge_8{}^tBu_8Br_2$ compared to that of a regular cubane.

Cubanes and related compounds

The cubane motif is well exemplified within compounds formed between the later elements of group 14 and Lewis bases. Synthetic approaches to the cubanes include use of mononuclear precursors. Condensation reactions between $^nBuSn(O)OH$ and phosphinic acids, R_2PO_2H, yield $[^nBuSn(O)(\mu\text{-}O_2PR_2)]_4$. $PhSnCl_3$ reacts with RC_2OM (M = Na, Ag) in the presence of water to give $[PhSn(O)(\mu\text{-}O_2CR]_6$ (see Fig. 3.31). Cubanes are also synthesized by the cycloaddition of preformed units which may or may not be identical (Eqn 5.86).

A useful synthetic route to a variety of cubanes of type $[ENR']_4$ is the reaction of the monocyclic compound $Me_2Si(NR)_2E$ (R alkyl or aryl; E = Ge, Sn, Pb) with $R'NH_2$ (R' = alkyl or aryl) (Eqn 5.87). Product distribution is influenced by the steric properties of R'; bulky substituents favour cubanes whereas small R' groups may result in polymer rather than discrete cluster formation. The heavier elements in group 14 exhibit a preference for bond formation using pure *p*-orbitals; this is a consequence of the inert pair effect. The cubane structure, with 90° endocyclic bond angles, is well suited to this preference.

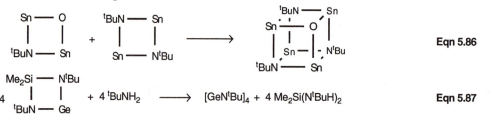

Eqn 5.86

$$4 \quad \begin{array}{c} Me_2Si \text{---} N^tBu \\ | \qquad | \\ {}^tBuN \text{---} Ge \end{array} \quad + 4\ {}^tBuNH_2 \longrightarrow [GeN^tBu]_4 + 4\ Me_2Si(N^tBuH)_2 \qquad \textbf{Eqn 5.87}$$

Adamantane-type clusters

One factor that determines the method adopted for the synthesis of an adamantane-type of cluster containing a germanium or tin atom is the role that the group 14 element plays in the cage (Fig. 5.16). In the first group with general formula $(RE)_4X_6$, each germanium or tin atom carries one *exo*-substituent. Typical syntheses are given in Eqns 5.88–5.92. In Eqn 5.89, H_2S can be replaced by H_2Se to yield $(PhGe)_4Se_6$. Similarly, the reaction may be adapted by using an organotin trihalide, e.g. $MeSnBr_3$, with sodium sulfide to prepare $(MeSn)_4S_6$. An alternative method of preparing $[Ge_4S_{10}]^{4-}$ is similar to the method used to synthesize $[In_4X_{10}]^{8-}$ (Eqn 5.65). $Tl_4[Ge_4S_{10}]$ can be isolated from a molten mixture of GeS_2 and Tl_2S; $Tl_4[Ge_4Se_{10}]$ is similarly prepared. The second group of clusters in Fig. 5.16 have the general formula $X_4(ER_2)_6$. Now the germanium or tin precursor needs to be di- rather than tri-functional, e.g. R_2SnCl_2 is suitable rather than $RSnCl_3$ (Eqn 5.93).

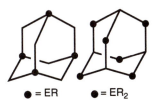

Fig. 5.16 The two classes of adamantane-type cluster containing Sn or Ge (E) atoms: $(RE)_4X_6$ and $X_4(ER_2)_6$.

$$4\ PhGeCl_3 + 6\ K_2PPh \longrightarrow (PhGe)_4(PPh)_6 + 12\ KCl \qquad \textbf{Eqn 5.88}$$

$$4\ PhGeCl_3 + 6\ H_2S \xrightarrow{Et_3N\ present} (PhGe)_4S_6 + 12\ HCl \qquad \textbf{Eqn 5.89}$$

$$4\ CF_3GeCl_3 + 6\ E(SiH_3)_2 \xrightarrow[E = S\ or\ Se]{Al_2Cl_6} (CF_3Ge)_4E_6 + 12\ SiH_3Cl \qquad \textbf{Eqn 5.90}$$

$$4\ GeBr_4 + 6\ H_2S \xrightarrow{Boiling\ CS_2} (BrGe)_4S_6 + 12\ HBr \qquad \textbf{Eqn 5.91}$$

$$4\ GeS_2 + 2\ S^{2-} \xrightarrow{Aq.;\ Cs^+\ present} [Ge_4S_{10}]^{4-} \qquad \textbf{Eqn 5.92}$$

$$6\ Ph_2SnCl_2 + 4\ PH_3 \longrightarrow P_4(SnPh_2)_6 + 12\ HCl \qquad \textbf{Eqn 5.93}$$

5.11 Phosphorus

Polycyclic phosphides and phosphanes

Syntheses of the anions $[P_7]^{3-}$, $[P_{16}]^{2-}$, $[P_{21}]^{3-}$, and $[P_{26}]^{4-}$ from white phosphorus were described in Section 2.3. Other methods of accessing the $[P_7]^{3-}$ anion are given in Eqns 5.94 and 5.95. The lithium salt Li_3P_7 is an important starting material for the synthesis of alkyl derivatives, P_7R_3 (Eqn 5.96). Of the two configurational isomers of P_7R_3 shown in Fig. 5.17, isomer (b) is favoured, as lone pair---lone pair repulsions (and R---R interactions) are minimized.

$$9\ P_2H_4 + 3\ ^nBuLi \xrightarrow{THF,\ -20°C} Li_3P_7 + 11\ PH_3 + 3\ ^nBuH \qquad \textbf{Eqn 5.94}$$

$$9\ P_2H_4 + 3\ LiPH_2 \xrightarrow{Monoglyme,\ -20°C} Li_3P_7 + 14\ PH_3 \qquad \textbf{Eqn 5.95}$$

$$Li_3P_7 + 3\ MeBr \longrightarrow P_7Me_3 + 3\ LiBr \qquad \textbf{Eqn 5.96}$$

$$n\ RPCl_2 + m\ PCl_3 \xrightarrow[-\ MgCl_2]{Mg} P_{n+m}R_n \qquad \textbf{Eqn 5.97}$$

$$n\ RPCl_2 + {}^m/_4\ P_4 \xrightarrow[-\ MgCl_2]{Mg} P_{n+m}R_n \qquad \textbf{Eqn 5.98}$$

$$n\ cyclo\text{-}[PR]_x + m\ PCl_3 \xrightarrow[-\ MgCl_2]{Mg} P_{xn+m}R_{xn} \qquad \textbf{Eqn 5.99}$$

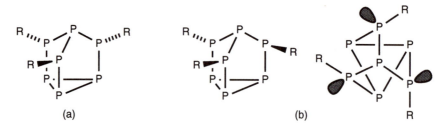

Fig. 5.17 The two configurational isomers of P_7R_3. Isomer (b) is favoured. In this isomer, the substituents are positioned such that they adopt a *paddle-wheel* motif and the lone pairs of electrons shown in the plan-view of isomer (b) are as far apart as possible.

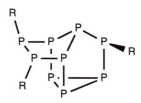

Fig. 5.18 Structure of P_9R_3.

Cyclopolyphosphanes P_nR_m can be prepared by the three methods shown in Eqns 5.97–5.99. The second method tends to yield products with a higher phosphorus content than the first or third routes. In these, products with between five and nine phosphorus atoms are favoured, in particular P_7R_3 and P_9R_3 (Fig. 5.18) The latter can be prepared from the former by a ring expansion reaction; Li_3P_7 reacts with ClRP–PRCl (e.g. R = ^tBu) and after treatment with RCl, P_9R_3 forms as a final equivalent of LiCl is eliminated.

Phosphorus-containing cubanes

A few phosphorus-containing cubane clusters have been mentioned in earlier sections, e.g. $[^tBuAlP(SiPh_3)]_4$ and $Cp^*_2Ti_2P_6$. The methods of synthesis of the group 14 cubanes (e.g. $[GeN^tBu]_4$ in Eqn 5.87) could reasonably be interpreted in terms of an intermediate species {ENR} but this is not an isolable starting material. On the other hand, for the phosphacubane $[^tBuCP]_4$, the monomeric unit $^tBuC{\equiv}P$ is indeed the precursor. When heated at 130°C for 65 hours, $^tBuC{\equiv}P$ undergoes a cyclotetramerization which is viewed as a stepwise process (Fig. 5.19). A better route to $[^tBuCP]_4$, is to treat $Cp_2Zr(^tBuC)_2P_2$ (Fig. 5.20) with C_2Cl_6; this removes the zirconium fragment and produces a $\{C_2P_2\}$-unit which readily dimerizes. $[^tBuCP]_4$, is an air stable, yellow crystalline solid.

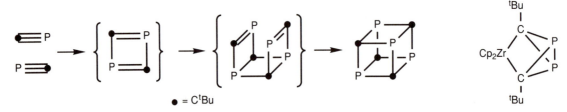

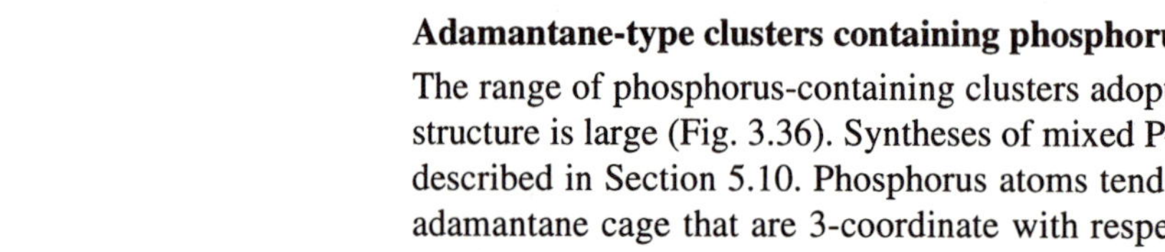

Fig. 5.19 The cyclotetramerization of ^tBuCP.

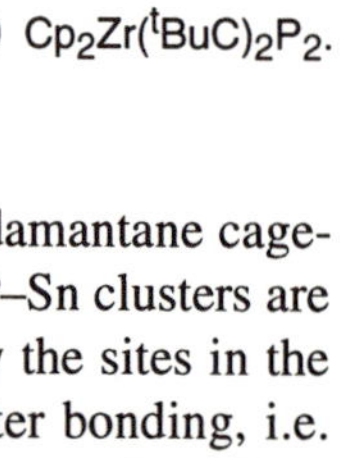

Fig. 5.20 $Cp_2Zr(^tBuC)_2P_2$.

Adamantane-type clusters containing phosphorus

The range of phosphorus-containing clusters adopting the adamantane cage-structure is large (Fig. 3.36). Syntheses of mixed P–Ge and P–Sn clusters are described in Section 5.10. Phosphorus atoms tend to occupy the sites in the adamantane cage that are 3-coordinate with respect to cluster bonding, i.e. the formulae of derivatives will be P_4Y_6 (where each P atom carries a lone pair of electrons) and P_4Y_{10} (where each P atoms bears an *exo*-substituent).

The reaction of PCl_3 with an excess of $MeNH_2$ proceeds by the elimination of HCl and the formation of $P_4(NMe)_6$. The $[P_4N_{10}]^{10-}$ anion (isoelectronic with P_4O_{10}) has been synthesized as the decalithium salt (Eqns 5.100–5.102).

$$4\,P_3N_5 + 10\,Li_3N \xrightarrow[\text{at } 720°C]{\text{in solid phase}} 3\,Li_{10}[P_4N_{10}] \qquad \textbf{Eqn 5.100}$$

$$4\,LiPN_2 + 2\,Li_3N \xrightarrow[\text{at } 700°C]{\text{in solid phase}} Li_{10}[P_4N_{10}] \qquad \textbf{Eqn 5.101}$$

$$10\,Li_7PN_4 + 6\,P_3N_5 \xrightarrow[\text{at } 630°C]{\text{in solid phase}} 7\,Li_{10}[P_4N_{10}] \qquad \textbf{Eqn 5.102}$$

Fig. 5.21 $P_4O_3S_6$.

The neutral sulfide P_4S_{10} forms when red phosphorus is heated with sulfur at 350–400°C. P_4S_{10} decomposes in water, and at its boiling point (514°C) it dissociates into sulfur, P_4S_7, and P_4S_3. Heating a mixture of P_4O_{10} and P_4S_{10} yields $P_4O_3S_6$ (Fig. 5.21). Synthetic routes to phosphorus sulfides are shown in Fig. 5.22. The product distribution obtained by heating mixtures of elemental sulfur and phosphorus is determined by the ratio of reactants. An important reagent for transformations between phosphorus sulfides is PPh_3. This is a desulfurizing agent and is used to degrade a cluster, e.g. P_4S_7 to P_4S_6 or P_4S_5. The selenide P_4Se_3 is prepared by heating red phosphorus and selenium; in air, P_4Se_3 decomposes liberating H_2Se. P_4Se_4 is synthesized by heating P_4Se_3 with selenium at 250–300°C; at 350°C, P_4Se_{10} is produced. P_4Se_{10} can also be obtained by the direct combination of the elements. The reaction of P_4Se_3 with bromine in CS_2 yields P_4Se_5.

Adamantane-like clusters exhibiting P–Si bonds fall into the two groups exemplified by $P_4(SiMe_2)_6$ and $P_7(SiMe_3)_3$. The latter is related to the polyphosphanes described earlier and is prepared from Li_3P_7 (Eqn 5.103). It is also a by-product of the reaction of white phosphorus with Me_2SiCl_2 in the presence of Na/K alloy, a route that is used to access $P_4(SiMe_2)_6$. $P_4(SiMe_2)_6$ is also formed according to Eqns 5.104 and 5.105. The *exo*-silyl groups in $P_7(SiMe_3)_3$ can be exchanged for $GeMe_3$, $SnMe_3$, or $PbMe_3$ substituents by treating $P_7(SiMe_3)_3$ with Me_3ECl (E = Ge, Sn, or Pb).

P_4S_{10} is an important *thiating agent*, being used to introduce sulfur into organic molecules (e.g. amides are converted to thioamides). It is also a deoxygenating agent, and is a precursor to organothiophosphorus compounds.

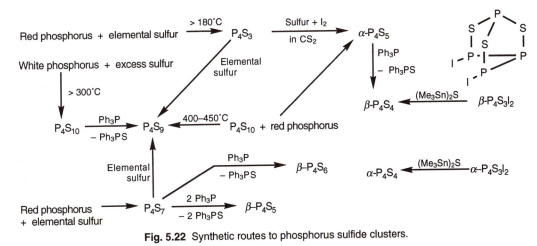

Fig. 5.22 Synthetic routes to phosphorus sulfide clusters.

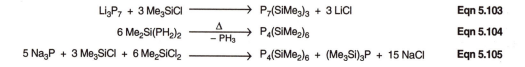

$$\text{Li}_3\text{P}_7 + 3\text{ Me}_3\text{SiCl} \longrightarrow \text{P}_7(\text{SiMe}_3)_3 + 3\text{ LiCl} \qquad \textbf{Eqn 5.103}$$

$$6\text{ Me}_2\text{Si}(\text{PH}_2)_2 \xrightarrow[-\text{PH}_3]{\Delta} \text{P}_4(\text{SiMe}_2)_6 \qquad \textbf{Eqn 5.104}$$

$$5\text{ Na}_3\text{P} + 3\text{ Me}_3\text{SiCl} + 6\text{ Me}_2\text{SiCl}_2 \longrightarrow \text{P}_4(\text{SiMe}_2)_6 + (\text{Me}_3\text{Si})_3\text{P} + 15\text{ NaCl} \qquad \textbf{Eqn 5.105}$$

5.12 Arsenic and antimony

The $[\text{As}_7]^{3-}$ anion, isostructural with $[\text{P}_7]^{3-}$, is prepared by heating together elemental arsenic and barium at 800°C; black crystals of $\text{Ba}_3[\text{As}_7]_2$ result. Alkali metal salts are useful precursors for derivatizing $[\text{As}_7]^{3-}$ and a reaction analogous to that in Eqn 5.103 can be used to form $\text{As}_7(\text{SiMe}_3)_3$. $\text{MeC}(\text{CH}_2)_3\text{As}_3$ is related to $[\text{As}_7]^{3-}$ and is synthesized according to Eqn 5.106. On prolonged contact with the sodium in the reaction mixture, a reactive sodium arsenide is formed *in situ* and this reacts with oxygen (or abstracts it from the THF solvent) to give $\text{MeC}(\text{CH}_2)_3\text{As}_3\text{O}_3$ (Fig. 3.39). The reaction of $\text{MeC}(\text{CH}_2\text{AsI}_2)_3$ with RNH_2 yields $\text{MeC}(\text{CH}_2)_3\text{As}_3(\text{NR})_3$ with concomitant loss of HI. A reaction similar to that in Eqn 5.106 is used to prepare $\text{MeC}(\text{CH}_2)_3\text{Sb}_3$.

The adamantane-like cluster As_4O_{10} results from the combustion of α-arsenic in pure oxygen. As_4O_6 is also produced; the latter is formed in the absence of the higher oxide by heating the element in air.

The preparation of $\text{As}_4(\text{SiMe}_2)_6$ is a stepwise process (Eqn 5.107). The clusters $\text{As}_4(\text{NR})_6$ (R = Me, ^{i}Pr, ^{n}Bu) are prepared from AsCl_3 and the appropriate primary amine; the method is similar to that used to synthesize $\text{P}_4(\text{NMe})_6$. Similarly, $\text{Sb}_4(\text{NAr})_6$ (Ar = aryl) is synthesized by treating SbX_3 (X = I or OEt) with ArNH_2.

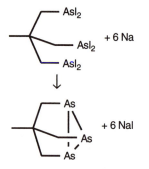

Eqn 5.106

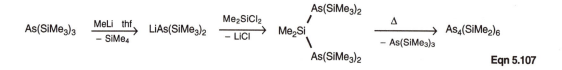

Eqn 5.107

Entry into the arsenic sulfides and selenides is gained by heating together elemental arsenic and sulfur or selenium. In both cases, As_4E_3 (E = S or Se) is formed. As_4S_3 can be converted into As_4S_4 by reaction with AsI_3; whether α- or β-As_4S_4 is formed depends on purification methods. Synthetic strategies here do *not* parallel those used in phosphorus sulfide chemistry.

5.13 Bismuth

The smallest of the cationic clusters of bismuth, $[\text{Bi}_5]^{3+}$, is produced when BiCl_3 is reduced by elemental bismuth at 250–270°C (molten conditions) in the presence of a Lewis acid (e.g. AlCl_3) and MAlCl_4 (M = Na or K). The cation is isolated as its $[\text{AlCl}_4]^-$ salt. The $[\text{Bi}_8]^{2+}$ ion is also produced and the reaction can be swung in its favour by altering the stoichiometry. A second route to $[\text{Bi}_5]^{3+}$ is shown in Eqn 5.108. Syntheses for $[\text{Bi}_9]^{5+}$ are given in Eqns 5.109 and 5.110; the systems are molten.

$$\text{Bi} + \text{AsF}_5 \xrightarrow{\text{Liquid SO}_2} [\text{Bi}_5][\text{AsF}_6]_3 + \text{AsF}_3 + \text{other products} \qquad \textbf{Eqn 5.108}$$

$$28\,\text{BiCl}_3 + 44\,\text{Bi} \xrightarrow[\approx 300^\circ\text{C}]{\text{KCl–BiCl}_3 \text{ solvent}} 3\,\text{Bi}_{24}\text{Cl}_{28} = [\text{Bi}_9^{5+}]_2[\text{BiCl}_5^{2-}]_4[\text{Bi}_2\text{Cl}_8^{2-}] \qquad \textbf{Eqn 5.109}$$

$$2\,\text{BiCl}_3 + 3\,\text{HfCl}_4 + 8\,\text{Bi} \xrightarrow{450^\circ\text{C}} [\text{Bi}_9^{5+}][\text{Bi}^+][\text{HfCl}_6^{2-}]_3 \qquad \textbf{Eqn 5.110}$$

5.14 Sulfur, selenium, and tellurium

S_4N_4 and related compounds

Tetrasulfur tetranitride, S_4N_4, is synthesized from S_2Cl_2 (Eqns 5.111 and 5.112) and must be handled with care; it explodes when heated or struck, and pure samples are particularly sensitive. Se_4N_4 is prepared either from $SeCl_4$ or $(Et_2O)_2SeO$ with ammonia; it is an orange, hygroscopic solid and, like S_4N_4, it is explosive.

$$6\,S_2Cl_2 + 16\,NH_3 \xrightarrow{50^\circ\text{C}} S_4N_4 + 8\,S + 12\,NH_4Cl \qquad \textbf{Eqn 5.111}$$

$$6\,S_2Cl_2 + 4\,NH_4Cl \xrightarrow{160^\circ\text{C}} S_4N_4 + 8\,S + 16\,HCl \qquad \textbf{Eqn 5.112}$$

The cluster As_4S_4 was mentioned briefly in Section 5.12; α- and β-As_4S_4 may be formed from As_4S_3 (see above). α-As_4S_4 occurs as the mineral *realgar*. At 270°C, realgar transforms to β-As_4S_4. If arsenic and sulfur are heated together at 500–600°C and the product is cooled rapidly, only the β-form results. Black crystals of As_4Se_4 are formed after pyrolyzing the constituent elements at 500°C. The product must be annealed at 250°C since decomposition to As_2Se_3 occurs at 264°C.

Cyclic sulfur, selenium, and tellurium cations

The oxidation of the sulfur, selenium, or tellurium with arsenic or antimony pentafluorides yields cationic species (Eqns 5.113–5.115). Although these systems are actually mono- or bicyclic rings, they are sometimes considered to be cluster compounds. On the other hand, $[Te_6]^{4+}$ is a true cluster and is formed according to Eqn 5.116.

AsF_5 and SbF_5 are both oxidizing agents and halide acceptors:
Reduction:
$AsF_5 + 2e^- \rightarrow AsF_3 + 2\,F^-$
$SbF_5 + 2e^- \rightarrow SbF_3 + 2\,F^-$
Halide acceptor:
$AsF_5 + F^- \rightarrow [AsF_6]^-$
$SbF_5 + F^- \rightarrow [SbF_6]^-$
$SbF_5 + SbF_6^- \rightarrow [Sb_2F_{11}]^-$

$$S_8 + 3\,AsF_5 \longrightarrow [S_8][AsF_6]_2 + AsF_3 \qquad \textbf{Eqn 5.113}$$

$$Se_8 + 5\,SbF_5 \xrightarrow{SO_2,\,-23^\circ\text{C}} [Se_8][Sb_2F_{11}]_2 + SbF_3 \qquad \textbf{Eqn 5.114}$$

$$Se_8 + 6\,AsF_5 \xrightarrow{SO_2,\,8^\circ\text{C}} 2\,[Se_4][AsF_6]_2 + 2\,AsF_3 \qquad \textbf{Eqn 5.115}$$

$$6\,Te + 6\,AsF_5 \xrightarrow[-196^\circ\text{C}]{AsF_3 \text{ or } SO_2} [Te_6][AsF_6]_4 + 2\,AsF_3 \qquad \textbf{Eqn 5.116}$$

6   Reactivity of clusters

6.1   Introduction

Although many clusters containing main group elements have been synthesized and structurally characterized, there remains much to learn by way of their reactivities. This chapter focuses on those groups of clusters of the p-block elements for which reactivity studies have been carried out. The data are selective and the aim of the chapter is to provide the reader with some insight into typical chemical properties of the cluster compounds.

6.2   Borane clusters

Certain aspects of the reactivity of borane clusters have already been explored during the discussion of syntheses in Chapter 5, for example cage expansion, cage coupling, cage fusion, and the introduction of heteroatoms when the heteroatom is one from the p-block. A complete discussion of the reactivity of all known boranes is beyond the scope of this book and this section is concerned only with typical reactions of representative *closo*-, *nido*-, and *arachno*-boranes and hydroborate anions.

Closo-hydroborate dianions

Of the dianions in the series *closo*-$[B_nH_n]^{2-}$, (n = 6 to 12), $[B_{10}H_{10}]^{2-}$ and $[B_{12}H_{12}]^{2-}$ have received most attention. Along with $[B_6H_6]^{2-}$, they are the most thermally stable anions in the series although the thermal stability depends upon the counter-ion; $Ag_2[B_6H_6]$ detonates upon heating whereas $Cs_2[B_6H_6]$ is thermally stable to about 600°C. Both $[B_{10}H_{10}]^{2-}$ and $[B_{12}H_{12}]^{2-}$ are hydrolytically stable and are kinetically stable in aqueous acidic and alkaline solutions. The smaller *closo*-hydroborate dianions are significantly more susceptible to hydrolysis than are $[B_{10}H_{10}]^{2-}$ and $[B_{12}H_{12}]^{2-}$ and are more easily oxidized.

Given the delocalized nature of the bonding in borane clusters (see Section 4.6) it is not surprising that there are some similarities between the chemistry of the *closo*-hydroborate dianions and aromatic organic molecules; e.g. they undergo electrophilic substitution reactions. In icosahedral $[B_{12}H_{12}]^{2-}$ all the boron atoms are equivalent and the cluster bonding electrons are distributed evenly over the B_{12}-cage. Aspects of the chemical behaviour of $[B_{12}H_{12}]^{2-}$ resemble those of benzene. In electrophilic substitution reactions of $[B_{12}H_{12}]^{2-}$, there can be no site preference for the first substitution since all boron atoms are equivalent. In $[B_{10}H_{10}]^{2-}$, the apical atoms (sites 1 and 10 in Fig. 6.1) are more negatively charged than the eight equatorial boron atoms and are the first to be attacked by electrophiles.

A B–H bond is prone to hydrolysis according to the equation:

$$B–H + H_2O \rightarrow B–O–H + H_2$$

In each of the anions $[B_{12}H_{12}]^{2-}$ and $[B_6H_6]^{2-}$ the cage boron atoms are equivalent to one another; in all the other *closo*-hydroborate dianions, there are at least two sites that differ in terms of the connectivity number within the cage (see Fig. 3.1).

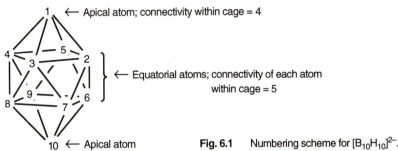

1 ← Apical atom; connectivity within cage = 4

} ← Equatorial atoms; connectivity of each atom within cage = 5

10 ← Apical atom

Fig. 6.1 Numbering scheme for $[B_{10}H_{10}]^{2-}$.

The net atomic charge centred at a boron atom in a *closo*-hydroborate cluster depends on the connectivity of the boron atom; sites of low connectivity are more negatively charged than sites of high connectivity.

The **connectivity** of an atom in a cluster is defined here as the number of neighbouring atoms *within* the cluster core.

Some reactions of $[B_{12}H_{12}]^{2-}$ are summarized in Fig. 6.2. For a di-substituted product, three isomers are possible. These are the 1,2-, 1,7-, and 1,12-isomers although all three may not necessarily be formed in a given substitution reaction. Carbonylation leads to a mixture of 1,7- and 1,12-$(CO)_2B_{12}H_{10}$ but in the presence of a dicobalt octacarbonyl catalyst, the 1,12-isomer is formed preferentially. The $[B_{12}H_{12}]^{2-}$ dianion reacts with a variety of bases (e.g. nitriles, nitrobenzene, sulfones, and sulfonamides) but only in strongly acidic solution. The stability of the cluster is apparent from its resistance to oxidation. In acetonitrile however, electrochemical oxidation occurs to give $[B_{24}H_{23}]^{3-}$.

Remember that once substituents are attached to the cage, the charge distribution over the boron cluster is perturbed.

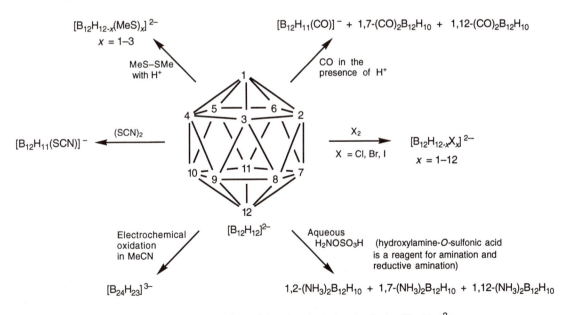

$[B_{12}H_{12-x}(MeS)_x]^{2-}$
$x = 1–3$

MeS–SMe with H^+

$[B_{12}H_{11}(CO)]^-$ + 1,7-$(CO)_2B_{12}H_{10}$ + 1,12-$(CO)_2B_{12}H_{10}$

CO in the presence of H^+

$[B_{12}H_{11}(SCN)]^-$ ← $(SCN)_2$

X_2
X = Cl, Br, I

$[B_{12}H_{12-x}X_x]^{2-}$
$x = 1–12$

Electrochemical oxidation in MeCN

$[B_{12}H_{12}]^{2-}$

Aqueous H_2NOSO_3H (hydroxylamine-*O*-sulfonic acid is a reagent for amination and reductive amination)

$[B_{24}H_{23}]^{3-}$

1,2-$(NH_3)_2B_{12}H_{10}$ + 1,7-$(NH_3)_2B_{12}H_{10}$ + 1,12-$(NH_3)_2B_{12}H_{10}$

Fig. 6.2 Selected reactions of the *closo*-hydroborate dianion $[B_{12}H_{12}]^{2-}$.

Some aspects of the reactivity of $[B_{10}H_{10}]^{2-}$ and $[B_{12}H_{12}]^{2-}$ are similar. The former resists attack by nucleophiles but is fairly reactive with respect to electrophilic substitution. Both $[B_{10}H_{10}]^{2-}$ and $[B_{12}H_{12}]^{2-}$ may be halogenated directly with Cl_2, Br_2, or I_2 in aqueous or alcoholic solutions to give $[B_{10}H_{10-x}X_x]^{2-}$ ($x = 1$ to 10) and $[B_{12}H_{12-x}X_x]^{2-}$ ($x = 1$ to 12); the rate of halogenation is Cl > Br > I and the rate decreases as x increases.

Adducts of $[B_{10}H_{10}]^{2-}$ with Lewis bases may be formed indirectly as in the case of $1,10\text{-}(CO)_2B_{10}H_8$. The compound is accessed via the diazo derivative $1,10\text{-}(N_2)_2B_{10}H_8$ (Eqn 6.1). Two moles of dinitrogen have formally replaced two hydride (H^-) ligands. Since dinitrogen is an excellent leaving group, carbon monoxide is able to displace N_2. In water, $1,10\text{-}(CO)_2B_{10}H_8$ is hydroxylated according to Eqn 6.2. $1,10\text{-}(N_2)_2B_{10}H_8$ is a useful precursor to a range of compounds and reacts with ammonia, azide ion, nitriles, and hydroxide ion to give $B_{10}H_8L_2$ (L = nucleophile).

$$[B_{10}H_{10}]^{2-} \xrightarrow[\text{2. } [BH_4]^-]{\text{1. excess HNO}_2} 1,10\text{-}(N_2)_2B_{10}H_8 \xrightarrow[-N_2]{CO} 1,10\text{-}(CO)_2B_{10}H_8 \qquad \textbf{Eqn 6.1}$$

$$1,10\text{-}(CO)_2B_{10}H_8 + 2 H_2O \rightleftharpoons [1,10\text{-}(COOH)_2B_{10}H_8]^{2-} + 2 H^+ \qquad \textbf{Eqn 6.2}$$

Neutral *nido*-boranes: B_5H_9 and $B_{10}H_{14}$

Pentaborane(9) and decaborane(14) are two representative *nido*-boranes, one a small and one a large cluster. Selected but typical reactions of B_5H_9 are given in Fig. 6.3.

$$B_2H_6 + 6 H_2O \longrightarrow 2 B(OH)_3 + 6 H_2 \qquad \textbf{Eqn 6.3}$$
$$\text{Boric acid}$$

$$2\text{-MeB}_5H_8 + 14 ROH \longrightarrow 4 B(OR)_3 + MeB(OR)_2 + 11 H_2 \qquad \textbf{Eqn 6.4}$$

Complete hydrolysis of a borane is a method of analysis; the number of moles of H_2 produced per mole of borane allows the stoichiometry of the borane to be deduced.

Whereas diborane(6) is hydrolysed rapidly in cold water to dihydrogen and boric acid (Eqn 6.3), *nido*-B_5H_9 hydrolyses only slowly. With an alcohol, ROH (Fig. 6.3) hydrolysis is complete; note that the elimination of dihydrogen drives the reaction. In the case of the hydrolysis of an organically substituted borane such as 2-MeB$_5$H$_8$, hydrolysis yields *H_2 and not MeH* (Eqn 6.4). The B–Me bond remains intact.

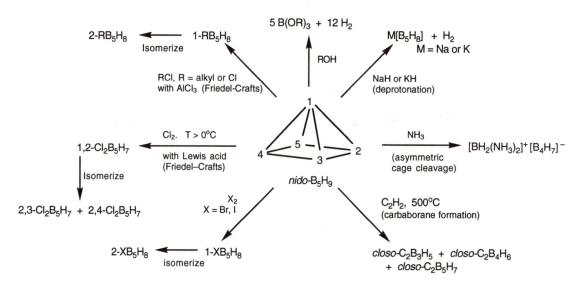

Fig. 6.3 Selected reactions of *nido*-B$_5$H$_9$.

Deprotonation of a *nido*-cluster is a simple but important reaction. The formation of an anionic hydroborate cluster provides a potential synthetic precursor, e.g. for metal-promoted cluster fusion (Section 5.1), the synthesis of heteroboranes (Sections 5.2 and 5.3), and routes to metallaborane clusters (see below). There are two types of hydrogen atom in B_5H_9, namely terminal and bridging atoms. A bridging hydrogen atom is removed upon treating B_5H_9 with a base such as potassium hydride. The preference for the loss of a bridge rather than terminal proton is a general observation for *nido*- and *arachno*-boranes and may be understood in terms of electron distribution (Fig. 6.4). Removal of a terminal proton would leave a boron-centred lone pair of electrons pointing out from the cluster. On the other hand, removal of a proton from a bridging site allows the electrons which were associated with the B–H–B bridge to be used in a direct B–B bonding interaction.

Deprotonation of a *nido*- or *arachno*-borane occurs by the loss of a bridging and not a terminal hydrogen atom.

Alkali metal hydrides are powerful deprotonating agents:
$$H^- + H^+ \rightarrow H_2$$

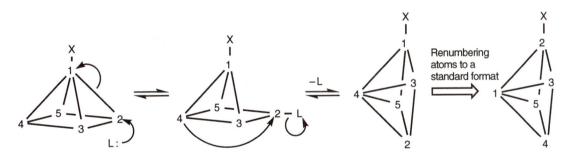

Deprotonation via the loss of a terminal hydrogen atom leaves a localized lone pair of electrons on the boron atom. This is not favourable.

Deprotonation via the loss of a bridging hydrogen atom allows a 3-centre-2-electron bond to become a 2-centre-2-electron bond.

Fig. 6.4 The effect on the localized electron distribution of removing either a terminal or bridging proton from a borane cluster.

Reactions of B_5H_9 with electrophiles give apically substituted products which isomerize to basally substituted clusters. The apical boron atom is more negatively charged than the basal atoms and so attracts an incoming electrophile. The isomerization of $1\text{-}XB_5H_8$ to $2\text{-}XB_5H_8$ does not involve the cleavage of a B–X bond; [10]B labelling studies show that the B_5-cage itself rearranges. One possible pathway is the *base-swing mechanism* (Fig. 6.5).

Fig. 6.5 The base-swing mechanism for the isomerization of $1\text{-}XB_5H_8$ to $2\text{-}XB_5H_8$. The process if catalysed by a base, L.

Whereas electrophiles attack the apical boron atom in pentaborane(9), Lewis bases attack the basal atoms. Donation of a lone pair of electrons from a Lewis base L may cause B–B bond cleavage as in Fig. 6.5 or can result in the opening of a bridging B–H–B interaction. Other boranes undergo similar reactions and for simplicity the mechanism is exemplified in Fig. 6.6 with the cleavage of B_2H_6 by a small Lewis base, L. During the cleavage, electrons which were associated with two 3-centre-2-electron bridge bonds become localized in new terminal B–H bonds. This type of reaction, the *asymmetric cleavage*, is characterized by the formation of a 1:1 electrolyte.

Fig. 6.6 The asymmetric cleavage of B_2H_6 by a sterically undemanding Lewis base, L. A similar mechanism applies to the asymmetric cleavage of B_5H_9 by ammonia to give $[B_4H_7]^-$ $[(NH_3)_2BH_2]^+$. See also Fig. 6.9.

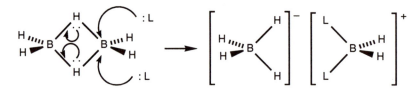

The reaction of B_5H_9 with acetylene at 500°C gives a mixture of *closo*-carboranes. With milder conditions (200°C), *nido*-$C_2B_4H_8$ is produced.

The reactivity of decaborane(14), an exemplary *large nido*-borane, differs significantly from that of B_5H_9; typical reactions are summarized in Fig. 6.7. $B_{10}H_{14}$ resists hydrolysis in neutral aqueous solution and is quite stable in air. Cage degradation occurs when $B_{10}H_{14}$ is treated with methanol in the presence of iodine. As with B_5H_9, the deprotonation of $B_{10}H_{14}$ proceeds by the abstraction of a proton from a bridging site. Up to two protons can be removed by hydride ion to give *nido*-$[B_{10}H_{13}]^-$ and *nido*-$[B_{10}H_{12}]^{2-}$.

In $B_{10}H_{14}$, cluster edges B(5)-B(6), B(6)-B(7), B(8)-B(9), and B(9)-B(10) are bridged by hydrogen atoms; atom numbering is given in Fig. 6.7.

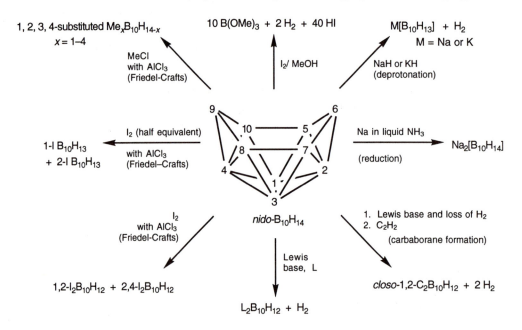

Fig. 6.7 Selected reactions of *nido*-$B_{10}H_{14}$.

The charge distribution in $B_{10}H_{14}$ renders atoms 1, 2, 3, and 4 the most negative and thus the atoms most susceptible to electrophilic attack. Atoms 6 and 9 are particularly susceptible to nucleophilic attack. Remember that the charge distribution over the cage is perturbed once substituents are attached.

Decaborane(14) undergoes both electrophilic and nucleophilic substitution reactions. Friedel–Crafts electrophilic substitutions give, for example, $Me_xB_{10}H_{14-x}$ ($x = 1$–4) with the Me groups in the 1, 2, 3, and 4-sites. Reaction of $B_{10}H_{14}$ with Me^- (e.g. LiMe) yields $Me_xB_{10}H_{14-x}$ ($x = 1$–4) with the Me groups in the 5, 6, 8, and 9-sites. Reacting $B_{10}H_{14}$ with the Lewis base Me_2S leads to the adduct 6,9-$(Me_2S)_2B_{10}H_{12}$; the ligands are easily replaced by other Lewis bases such as MeCN, Et_3N, and Ph_3P and by alkynes to give *closo*-carbaboranes $1,2-R_2C_2H_{10}H_{10}$. Unlike the small *nido*-cluster B_5H_9, $B_{10}H_{14}$ is not degraded by these Lewis bases.

The decaborane cage can be expanded either in a homo- or hetero-nuclear sense. Under mild conditions, $[BH_4]^-$ reduces $B_{10}H_{14}$ to the corresponding dianion (Eqn 6.5) but at 90°C, the product is *nido*-$[B_{11}H_{14}]^-$. Cluster fusion to give $B_{20}H_{16}$ occurs when a 1700V a.c. discharge is passed through a mixture of $B_{10}H_{14}$ vapour and dihydrogen.

$$\textit{nido-} \quad \xrightarrow{[BH_4]^-} \quad \textit{arachno}$$
$$B_{10}H_{14} \qquad\qquad [B_{10}H_{14}]^{2-}$$

Eqn 6.5

An *arachno*-borane: B_4H_{10}

Typical reactions of *arachno*-tetraborane(10) are given in Fig. 6.8. The hydrolysis of this small *arachno*-cluster is more facile than for either of the *nido*-species described above. Deprotonation follows the same pattern as for the *nido*-clusters with a proton being removed from a bridging position.

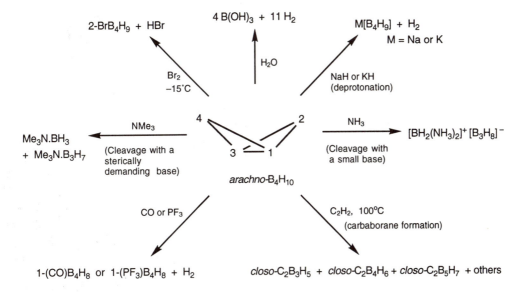

Fig. 6.8 Selected reactions of *arachno*-B_4H_{10}.

Amongst the reactions of B_4H_{10}, attack by nucleophiles features prominently. With a small base such as ammonia, the mechanism mimics that shown in Fig. 6.6 and the product is a 1:1 electrolyte. This pathway depends upon the two equivalents of the base being able to approach a single boron atom and is sterically hindered even with Lewis bases such as trimethylamine. Now the mechanism involves attack at two different boron sites with the formation of neutral aminoborane adducts (Fig. 6.9).

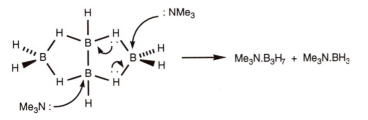

Fig. 6.9 Cleavage of B_4H_{10} by a sterically demanding Lewis base. During the process, the electrons from two 3-centre-2-electron B–H–B bridges are localized into two new B–H terminal bonds.

The reaction of *arachno*-B_4H_{10} with CO or PF_3 (π-acceptor ligands) proceeds via a substitution pathway. Dihydrogen is eliminated as the 2-electron donor molecule enters and hence the cluster electron count remains the same. Reactions with a Lewis bases possessing an active hydrogen atom (e.g. Me_2NH) can result in the elimination of H_2, cluster degradation, and the introduction of a bridging group derived from the base (Eqn 6.6).

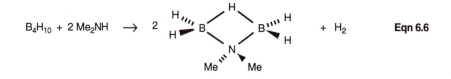

$$B_4H_{10} + 2\,Me_2NH \longrightarrow 2 \quad\quad\quad\quad + H_2 \quad\quad \text{Eqn 6.6}$$

6.2 *Nido-* and *arachno*-hydroborate anions and carbaborate anions as precursors to metallaboranes and metallacarbaboranes

Isobality is defined in Section 4.6.

Reactions of some *nido-* and *arachno*-hydroborate anions are described in Section 5.1. These include uses of the anions as precursors to higher nuclearity boranes. Open cage hydroborate anions are important precursors to heteroboranes, and syntheses of those incorporating *p*-block elements are described in Section 5.3. Reactions between *nido-* and *arachno*-hydroborate anions and suitable transition metal fragments yield a range of metallaborane clusters. Whether or not the transition metal group is fully or partially incorporated into the cluster depends upon the compatability of the frontier orbitals of the metal unit and those of the anionic cluster. For example, an $\{AuL\}^+$ (L = phosphine) fragment possesses a low lying unoccupied MO of a_1 symmetry and is isolobal with a proton; thus the reaction of a *nido*-hydroborate monoanion with LAuCl leads to an auraborane in which the gold(I) phosphine fragment occupies a bridging site. The structure of the auraborane is derived from the parent neutral *nido*-borane (Eqn 6.7 and Fig. 6.10; compare the structure in Fig. 6.10 with that of $B_{10}H_{14}$ in Fig. 3.1).

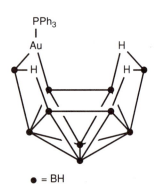

Fig. 6.10 Structure of $B_{10}H_{13}(AuPPh_3)$.

$$\text{*nido*-}[B_{10}H_{13}]^- + Ph_3PAuCl \longrightarrow \text{*nido*-}B_{10}H_{13}(AuPPh_3) + Cl^- \quad\quad \text{Eqn 6.7}$$

If the frontier orbitals of a transition metal fragment $\{ML_x\}$ mimic those of a $\{BH\}$-unit, then it should be possible to replace a $\{BH\}$-unit in a borane cluster by an $\{ML_x\}$-unit. The structure of the metallaborane product should resemble that of the borane precursor. Addition of $\{ML_x\}$-fragments to boranes and hydroborate anions can also occur. Two idealized reactions are shown in Eqns 6.8 and 6.9; in each the $\{ML_x\}$-unit provides two valence electrons. In the first case, the *nido*-framework is retained because the cluster electron count does not change during the substitution. In the second, a metal fragment adds to the borane; H_2 is lost to preserve the cluster electron count.

$$\text{*nido*-}B_nH_{n+4} + \{ML_x\} \longrightarrow \text{*nido*-}B_{n-1}H_{n+3}ML_x + \{BH\} \quad\quad \text{Eqn 6.8}$$

$$\text{*nido*-}B_nH_{n+4} + \{ML_x\} \longrightarrow \text{*closo*-}B_nH_{n+2}ML_x + H_2 \quad\quad \text{Eqn 6.9}$$

In practice, reactions are not always as straightforward as they might appear to be in theory; e.g. H_2 or ligands may be lost during the reaction and this alters the cluster electron count. Examples are given in Eqns 6.10–6.12. In Eqn 6.10, the interaction of the pentagonal open face of *nido*-$[C_2B_9H_{11}]^{2-}$ with an iron(II) centre results in a closed cluster with an FeC_2B_9-core. The final product is a *commo*-cluster; compare Eqn 6.10 with Eqn 5.34 and compare the *commo*-cluster with the structure of ferrocene in Fig. 6.11. A related reaction occurs between *nido*-$[B_{11}H_{13}]^{2-}$ and $(\eta^5\text{-}C_5H_5)_2Ni$ (Eqn 6.11). The latter is a source of an $\{(\eta^5\text{-}C_5H_5)Ni\}$-fragment and this provides three electrons for cluster bonding. During the reaction, H_2 is lost and this allows the product to retain the correct number of cluster bonding electrons to be a *closo*-cage. When *nido*-$[B_5H_8]^-$ reacts with *trans*-$Ir(PPh_3)_2(CO)Cl$, chloride ion is lost and the $\{Ir(PPh_3)_2(CO)\}^+$-fragment adds to the hydroborate anion, this time *without* loss of H_2 (Eqn 6.12). Since $\{Ir(PPh_3)_2(CO)\}^+$ provides two electrons to cluster bonding, the product is still a *nido*-cage; whereas $[B_5H_8]^-$ is a square based pyramidal cluster, the iridaborane is a pentagonal pyramid.

Refer to Section 4.6 for methods of determining the number of electrons provided by a cluster fragment for cluster bonding.

Refer to Section 4.6 for derivation of cluster shapes.

$$nido\text{-}[C_2B_9H_{11}]^{2-} + FeCl_2 \xrightarrow[-2\ Cl^-]{} [commo\text{-}3,3'\text{-}Fe(3\text{-}Fe\text{-}1,2\text{-}C_2B_9H_{11})_2]^{2-} \qquad \textbf{Eqn 6.10}$$

$$nido\text{-}[B_{11}H_{13}]^{2-} + (\eta^5\text{-}C_5H_5)_2Ni \xrightarrow[-H_2]{MeCN,\ Na/Hg} nido\text{-}[1\text{-}(\eta^5\text{-}C_5H_5)NiB_{11}H_{11}]^- \qquad \textbf{Eqn 6.11}$$

$$nido\text{-}[B_5H_8]^- + trans\text{-}Ir(PPh_3)_2(CO)(Cl) \xrightarrow[-Cl^-]{} nido\text{-}2\text{-}Ir(PPh_3)_2(CO)B_5H_8 \qquad \textbf{Eqn 6.12}$$

The *nido*-clusters 2,3-R_2-2,3-$C_2B_4H_6$ (R = H, alkyl, aryl, or SiMe₃) have been used as precursors to a series of *stacked compounds* which are related to organometallic *sandwich compounds*. The reaction shown in Fig. 6.11a results in the capping of the carbaborane cage with a $\{CpCo\}$-fragment. By *PSEPT*, the product may be considered as a *closo*-cobaltacarbaborane but another description is to liken it to ferrocene (Fig. 6.11b). The open face of the C_2B_4-cage and the cyclopentadienyl ring provide comparable sets of frontier orbitals. The carbaborate anion can therefore mimic an $\eta^5\text{-}C_5H_5$ organic π-ligand. The same is true of the *nido*-$[C_2B_9H_{11}]^{2-}$ anion described above.

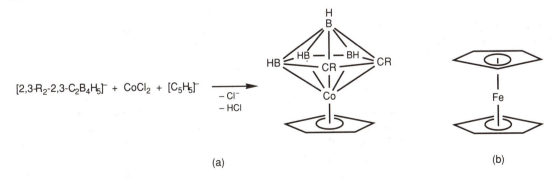

(a) (b)

Fig. 6.11 (a) Capping a *nido*-carbaborate anion to give a *closo*-metallacarbaborane. The product can be compared with an organometallic π-complex such as ferrocene, (b).

[(η⁶-C₆H₆)RuCl₂]₂ is a source of the {Ru(η⁶-C₆H₆)²⁺} fragment.

Deprotonation of *nido*-2,3-(SiMe₃)₂-2,3-C₂B₄H₆ with an excess of ⁿBuLi gives the dianion [*nido*-2,3-(SiMe₃)₂-2,3-C₂B₄H₄]²⁻. The open face of the cluster is capped by a ruthenium fragment when it is treated with half an equivalent of [(η⁶-C₆H₆)RuCl₂]₂, (Fig. 6.12). It is possible to *decap* the ruthenacarbaborane cluster thereby generating a *nido*-cluster that can be capped again with ruthenium to give an extended stack. This general sequence of events is the basis for a range of reactions providing routes to stacked systems.

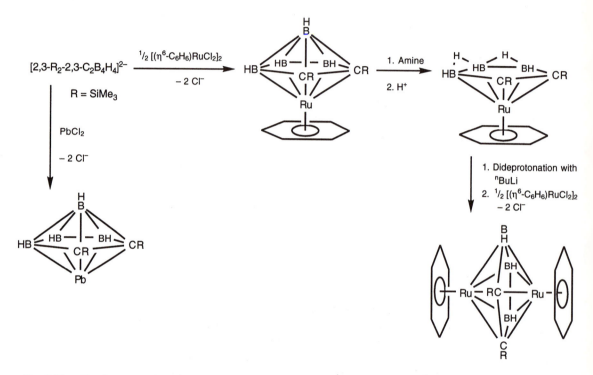

Fig. 6.12 The formation of *closo*-metallacarbaboranes from *nido*-[2,3-R₂-2,3-C₂B₄H₄]²⁻ (R = SiMe₃). Extension of the reaction sequence can yield molecules with stacked ring-units.

6.3 The carbaborane C₂B₁₀H₁₂

The use of *nido*-[1,2-C₂B₉H₁₁]²⁻ as a precursor to metallacarbaboranes is described in the previous section.

Of all carbaborane clusters, C₂B₁₀H₁₂ has been the most thoroughly studied; it is isoelectronic with [B₁₂H₁₂]²⁻. Of the three isomers (Fig. 3.6), 1,2-C₂B₁₀H₁₂ has received the most attention. 1,2-C₂B₁₀H₁₂ has a similar pseudo-aromatic character to [B₁₂H₁₂]²⁻. It resists attack by oxidizing agents, alcohols, and strong acids and is thermally stable to 400°C. Selected reactions of 1,2-C₂B₁₀H₁₂ are given in Fig. 6.13. Electrophilic substitution reactions at boron sites occur without cage degradation. Selective degradation of 1,2-C₂B₁₀H₁₂ via the extrusion of one boron-vertex occurs when the cluster is treated with methoxide ion in methanol. This reaction gives a route to the *nido*-[1,2-C₂B₉H₁₂]⁻ anion and, after deprotonation (Fig. 6.13), to *nido*-[1,2-C₂B₉H₁₁]²⁻.

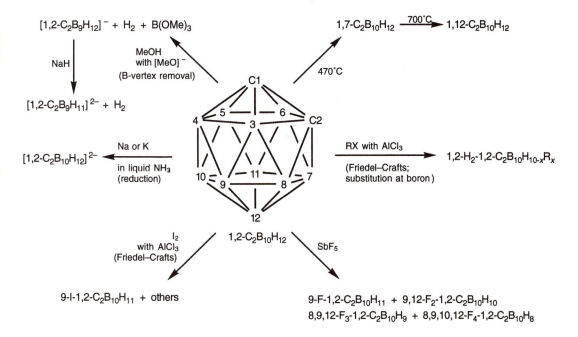

$[1,2\text{-}C_2B_9H_{12}]^- + H_2 + B(OMe)_3$

$\downarrow$ NaH

$[1,2\text{-}C_2B_9H_{11}]^{2-} + H_2$

MeOH
with $[MeO]^-$
(B-vertex removal)

$1,7\text{-}C_2B_{10}H_{12}$ $\xrightarrow{700°C}$ $1,12\text{-}C_2B_{10}H_{12}$

470°C

$[1,2\text{-}C_2B_{10}H_{12}]^{2-}$ $\xleftarrow[\substack{\text{in liquid } NH_3 \\ \text{(reduction)}}]{\text{Na or K}}$

RX with $AlCl_3$
(Friedel–Crafts;
substitution at boron)
$\rightarrow 1,2\text{-}H_2\text{-}1,2\text{-}C_2B_{10}H_{10-x}R_x$

I_2
with $AlCl_3$
(Friedel–Crafts)

$1,2\text{-}C_2B_{10}H_{12}$

SbF_5

$9\text{-}I\text{-}1,2\text{-}C_2B_{10}H_{11}$ + others

$9\text{-}F\text{-}1,2\text{-}C_2B_{10}H_{11}$ + $9,12\text{-}F_2\text{-}1,2\text{-}C_2B_{10}H_{10}$
$8,9,12\text{-}F_3\text{-}1,2\text{-}C_2B_{10}H_9$ + $8,9,10,12\text{-}F_4\text{-}1,2\text{-}C_2B_{10}H_8$

Fig. 6.13   Selected reactions of *closo*-$1,2\text{-}C_2B_{10}H_{12}$.

The hydrogen atoms attached to the C_2-unit in $1,2\text{-}C_2B_{10}H_{12}$ are acidic and this property is exploited in metallation reactions (Eqn 6.13). Lithiated derivatives can be used to prepare *C*-substituted compounds.

$1,2\text{-}C_2B_{10}H_{12} + 2\,RLi \longrightarrow 1,2\text{-}Li_2\text{-}1,2\text{-}C_2B_{10}H_{10} + 2\,RH$ **Eqn 6.13**

$2\ 1\text{-}Li\text{-}1,2\text{-}C_2B_{10}H_{11} \rightleftharpoons 1,2\text{-}Li_2\text{-}1,2\text{-}C_2B_{10}H_{10} + 1,2\text{-}C_2B_{10}H_{12}$ **Eqn 6.14**

The mono-lithiated cluster is difficult to isolate as it undergoes disproportionation (Eqn 6.14). However, monosubstituted derivatives may be accessed by using a protecting group as shown in Eqn 6.15. The dilithiated cluster $1,2\text{-}Li_2\text{-}1,2\text{-}C_2B_{10}H_{10}$ is a precursor to di-*C*-substituted derivatives of the carbaborane cluster; e.g. reaction with iodine gives $1,2\text{-}I_2\text{-}1,2\text{-}C_2B_{10}H_{10}$, and reaction with NOCl gives $1,2\text{-}(NO)_2\text{-}1,2\text{-}C_2B_{10}H_{10}$. The controlled bromination of $1,2\text{-}Li_2\text{-}1,2\text{-}C_2B_{10}H_{10}$ gives a mono-brominated species. Elimination of LiBr yields an unstable unsaturated product that can be trapped by reactions with suitable dienes.

$1,2\text{-}R_2\text{-}1,2\text{-}C_2B_{10}H_{10}$
can be represented as:

$1,2\text{-}C_2B_{10}H_{12}$ $\xrightarrow[\substack{\text{2. }^tBuMe_2SiCl \\ \text{(protection group} \\ \text{introduced)}}]{\text{1. }^nBuLi}$ $1\text{-}SiMe_2{}^tBu\text{-}1,2\text{-}C_2B_{10}H_{11}$ $\xrightarrow[\substack{\text{2. RCl} \\ \text{R = organic} \\ \text{substituent}}]{\text{1. }^nBuLi}$ $1\text{-}SiMe_2{}^tBu\text{-}2\text{-}R\text{-}1,2\text{-}C_2B_{10}H_{10}$

$\downarrow$ $^nBu_4NF$ in THF (deprotection)

$2\text{-}R\text{-}1,2\text{-}C_2B_{10}H_{11}$

(or renumbering to a standard format
gives $1\text{-}R\text{-}1,2\text{-}C_2B_{10}H_{11}$)

Eqn 6.15

The incorporation of {1,2-C_2B_{10}}-cluster units into the backbone of polymers increases the thermal stability of the polymer and also the solubility of the material in organic solvents. Polymeric materials can be formed by reacting suitable alkynes with decaborane(14) in order to construct the {C_2B_{10}}-cages *in situ*. Alternatively, suitably derivatized carborane clusters may undergo chain-building reactions (Eqns 6.16–6.17).

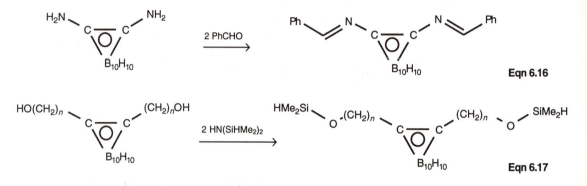

Eqn 6.16

Eqn 6.17

6.4 Boron halide clusters

The tetrahedral molecule B_4Cl_4 is relatively thermally stable; some reactions are summarized in Fig. 6.14. The halide substituents can be exchanged either by bromide ion or a bulky alkyl group. The conditions of the reaction with BBr_3 are critical; too low a temperature (T<97°C) gives no reaction whereas too high a temperature (T>103°C) leads to cluster fragmentation. Exchange of Cl^- for H^- is accompanied by cluster expansion. The reaction of B_4Cl_4 with diborane(6) proceeds via addition to give initially $B_6H_6Cl_4$ (structurally related to *nido*-B_6H_{10}). Further addition of B_2H_6 and exchange of chloride for hydrogen substituents leads to $B_{10}Cl_nH_{14-n}$ (n = 8–12) (related to *nido*-$B_{10}H_{14}$). Treatment of B_4Cl_4 with the reducing agent Me_3SnH opens the tetrahedral cluster to the butterfly structure of B_4H_{10}.

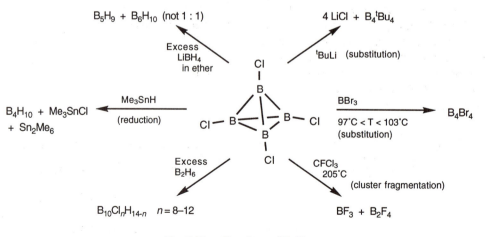

Fig. 6.14   Reactions of B_4Cl_4.

Fragmentation of the B_4-core of B_4Cl_4 into mono- and diboron species occurs when the cluster is treated with a variety of reagents. In the reaction with $CFCl_3$ (Fig. 6.14), the fragmentation is accompanied by exchange of the chloride for fluoride substituents. Similarly, Me_2NH reacts with B_4Cl_4 to give $(Me_2N)_2BCl$ and $B_2(NMe_2)_2Cl_2$ as well as an adduct, $B_2Cl_4.2HNMe_2$.

Expansion of the B_8-core of B_8Cl_8 accompanies most of its reactions. An exception is bromination (Eqn 6.18). The reaction of B_8Cl_8 with H_2 or diborane(6) at room temperature yields $B_9Cl_{9-n}H_n$ ($n = 0$–2); treatment of B_8Cl_8 with $AlMe_3$ gives $B_9Cl_{9-n}Me_n$ ($n = 0$–4). With ^tBuLi, substituent exchange occurs to give both $B_9{}^tBu_9$ and the reduced species $[B_9{}^tBu_9]^{2-}$. The cluster expansion is probably the result of disproportionation (Eqn 6.19) but the B_7-products are unstable. Preference for the B_9-cage is also observed amongst products from reactions of $B_{10}Cl_{10}$ (Eqn 6.20–6.22). Similar reactions occur for $B_{11}Cl_{11}$.

$$B_8Cl_8 \xrightarrow[\text{(Friedel–Crafts)}]{\text{BBr}_3 \text{ with AlCl}_3} B_8Br_8 \qquad \textbf{Eqn 6.18}$$

$$2\,B_8X_8 \longrightarrow \{B_7X_7\} + B_9X_9 \qquad \textbf{Eqn 6.19}$$

$$B_{10}Cl_{10} \xrightarrow{\text{Br}_2 \text{ or I}_2, \ 135^\circ C} B_9X_9 \quad (X = Br \text{ or } I) \qquad \textbf{Eqn 6.20}$$

$$B_{10}Cl_{10} \xrightarrow{\text{BBr}_3, \ 200^\circ C} B_9Br_9 \qquad \textbf{Eqn 6.21}$$

$$B_{10}Cl_{10} \xrightarrow{\text{H}_2, \ 150^\circ C} B_9Cl_8H \qquad \textbf{Eqn 6.22}$$

$$B_9Br_9 \xrightarrow{\text{SnMe}_4} B_9Br_{9-n}Me_n \quad (n = 1\text{-}9) \qquad \textbf{Eqn 6.23}$$

The halide cluster B_9Cl_9 is the most thermally stable member of the B_nX_n (X = halide) series and it is quite unreactive. Halide substitution occurs when B_9Cl_9 reacts with molten $AlBr_3$. In contrast to B_8Cl_8, B_9Cl_9 fails to react with dihydrogen. Alkyl derivatives of the B_9-cage can be prepared from B_9Br_9 (Eqn 6.23) or from B_8Cl_8 (see above). The bromide substituents in B_9Br_9 can be exchanged for chloride groups if the cluster is treated with $TiCl_4$ at $250^\circ C$.

6.5 Iminoalanes and related clusters

Iminoalanes (see Figs. 3.14 and 3.15) are air sensitive compounds and are rapidly hydrolysed by water. In clusters of the general type $[HAlNR]_n$, the exocyclic hydrogen atoms may be replaced by halogen atoms (Eqns 6.24–6.26). The reactions shown occur without cleavage of the cluster-core but this is not always the case; reaction of $[HAlN^iPr]_6$ with HCl or $HgCl_2$ results in iminoalane species other than $[ClAlN^iPr]_6$, and reaction of $[HAlN^nPr]_8$ with $TiCl_4$ yields not only $[ClAlN^nPr]_8$ but also $[ClAlN^nPr]_6$ and $[ClAlN^nPr]_{10}$.

$$[HAlNR]_n + n\,HCl \longrightarrow [ClAlNR]_n + n\,H_2 \qquad \textbf{Eqn 6.24}$$

$$2\,[HAlNR]_n + 2n\,TiCl_4 \longrightarrow 2\,[ClAlNR]_n + 2n\,TiCl_3 + n\,H_2 \qquad \textbf{Eqn 6.25}$$

$$2\,[HAlNR]_n + n\,HgCl_2 \longrightarrow 2\,[ClAlNR]_n + n\,Hg + n\,H_2 \qquad \textbf{Eqn 6.26}$$

Iminoalanes are used as reducing agents; $[HAlNR]_n$ (R = ^tBu, $n = 4$; R = ^iPr, $n = 6$; R = ^nPr, $n = 8$) clusters are used in the homogeneous catalytic hydrogenation of alkenes, and $[HAlN^iPr]_6$ reduces aldehydes and ketones to alcohols and facilitates the selective reduction of some dicarbonyl compounds. [HAlNR]$_n$ clusters are also used as polymerization catalysts.

The aluminaphosphacubane [^{i}BuAlP(SiPh$_3$)]$_4$ is susceptible to both electrophilic (at P) and nucleophilic (at Al) attack. Treatment with ethanol results in cluster cleavage with the formation of (Ph$_3$Si)PH$_2$ and ^{i}BuAl(OEt)$_2$; hydrolysis proceeds in a similar way. One potential application of clusters of this type is as precursors to solid-state materials and for aluminaphosphacubanes, the goal, after stripping away the exocyclic substituents, is aluminium phosphide.

6.6   Zintl ions with group 14 elements

Few studies concerning the reactivity of Zintl ions have been carried out; contrast this with the wealth of chemical information known about boranes to which Zintl ions are related both in terms of structure and bonding. A reaction that has a parallel in borane cluster chemistry is the transformation of a *nido-* to a *closo*-cage. The Zintl anions *nido-*[Sn$_9$]$^{4-}$ and *nido-*[Pb$_9$]$^{4-}$ (Fig. 3.26) react with (η^6-1,3,5-Me$_3$C$_6$H$_3$)Cr(CO)$_3$ to give *closo-*[Sn$_9$Cr(CO)$_3$]$^{4-}$ and *closo-*[Pb$_9$Cr(CO)$_3$]$^{4-}$ respectively (Fig. 6.15). The organic π-ligand is displaced by the Zintl ion and the result of the reaction is the incorporation of the chromium atom into the cluster. Since the {Cr(CO)$_3$}-fragment does not provide any valence electrons for cluster bonding (see Section 4.6), the *nido-*cluster is simply capped by the transition metal unit and undergoes no other structural perturbation.

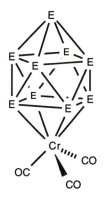

Fig. 6.15 Structure of *closo-*[E$_9$Cr(CO)$_3$]$^{4-}$ (E = Sn or Pb).

6.7   Cubanes other than iminoalanes

In general, a cubane contains donor and acceptor atoms. In [SnN^{t}Bu]$_4$, the tin atoms are acceptors and the ^{t}BuN-groups are donors with respect to the endocyclic bonding. Each tin atom still retains an exocyclic lone pair of electrons and is a potential Lewis base. The reactions in Eqns 6.27 and 6.28 illustrate this property. In a cubane with more than one donor site, preferential coordination to a Lewis acid can occur (Eqn 6.28).

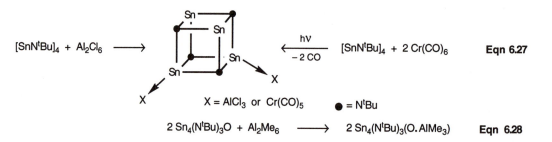

[SnN^{t}Bu]$_4$ + Al$_2$Cl$_6$ ⟶ ⟵ $\frac{h\nu}{-2\,CO}$ [SnN^{t}Bu]$_4$ + 2 Cr(CO)$_6$ **Eqn 6.27**

X = AlCl$_3$ or Cr(CO)$_5$ ● = N^{t}Bu

2 Sn$_4$(N^{t}Bu)$_3$O + Al$_2$Me$_6$ ⟶ 2 Sn$_4$(N^{t}Bu)$_3$(O.AlMe$_3$)   **Eqn 6.28**

6.8   Homoatomic anions from group 15 and their derivatives

Some syntheses involving Li$_3$P$_7$ are described in Section 5.11.

The compounds M$_3$P$_7$, M$_3$As$_7$, and M$_3$P$_{11}$ (M = alkali metal) should all react with RCl (R = various) to give substituted derivatives of the types R$_3$P$_7$,

R_3As_7, and R_3P_{11}. Indeed, examples of such reactions are known (Eqns 6.29–6.34) but in some cases the pathway of the reaction is more complicated that anticipated. For example, Li_3P_7 reacts with Ph_3SiCl in toluene according to Eqn 6.29 but in THF, other products are obtained. The reaction shown in Eqn 6.30 succeeds in toluene but in THF or by using H_3SiBr in place of H_3SiI, polymerization products are obtained.

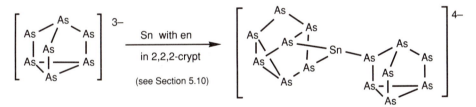

$$Li_3P_7 + 3\ Ph_3SiCl \xrightarrow{\text{toluene}} P_7(SiPh_3)_3 + 3\ LiCl \qquad \textbf{Eqn 6.29}$$

$$Li_3P_7 + 3\ H_3SiI \xrightarrow{\text{toluene}} P_7(SiH_3)_3 + 3\ LiI \qquad \textbf{Eqn 6.30}$$

$$Li_3P_7 + 3\ Me_3SnBr \xrightarrow{\text{toluene}} P_7(SnMe_3)_3 + 3\ LiBr \qquad \textbf{Eqn 6.31}$$

$$Na_3P_7 + 3\ Me_3GeCl \xrightarrow{\text{toluene}} P_7(GeMe_3)_3 + 3\ NaCl \qquad \textbf{Eqn 6.32}$$

$$Cs_3P_{11} + 3\ Me_3SiCl \longrightarrow P_{11}(SiMe_3)_3 + 3\ CsCl \qquad \textbf{Eqn 6.33}$$

$$Rb_3As_7 + 3\ Me_3SiCl \longrightarrow As_7(SiMe_3)_3 + 3\ RbCl \qquad \textbf{Eqn 6.34}$$

$$\underset{+\ 3\ Me_3SnCl}{P_7(SiMe_3)_3} \rightleftharpoons \underset{+\ 3\ Me_3SiCl}{P_7(SnMe_3)_3} \qquad \textbf{Eqn 6.35}$$

Equilibria can be established between related species (Eqn 6.35) but if the reaction is to be used as a method of exchanging the substituents, there must be a driving force to push the equilibrium to the right hand side. In Eqn 6.35, the low solubility of $P_7(SnMe_3)_3$ in the solvent dimethoxyethane allows precipitation of this product and its removal from the equilibrium system.

The oxidative coupling of two clusters is observed in the reaction of $[As_7]^{3-}$ with elemental tin (Fig. 6.16). When Rb_3As_7 is heated in 1,2-diaminoethane with $Fe_2(CO)_9$, the product is $[As_{22}]^{4-}$; this cluster can be viewed as resulting from the oxidative coupling of two $[As_{11}]^{3-}$ anions.

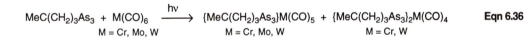

Fig. 6.16 Oxidative coupling of two $[As_7]^{3-}$ clusters.

$$MeC(CH_2)_3As_3 + M(CO)_6 \xrightarrow{h\nu} \{MeC(CH_2)_3As_3\}M(CO)_5 + \{MeC(CH_2)_3As_3\}_2M(CO)_4 \qquad \textbf{Eqn 6.36}$$
$$M = Cr, Mo, W \qquad\qquad M = Cr, Mo, W \qquad\qquad M = Cr, W$$

If a group 15 atom is incorporated into a cluster and still retains an exocyclic lone pair of electrons, the atom (and thus the cluster) is a potential Lewis base. $MeC(CH_2)_3As_3$ reacts with $Cr(CO)_6$, $Mo(CO)_6$, or $W(CO)_6$ according to Eqn 6.36 to give products in which the arsenic-containing cluster functions as a monodentate ligand. The cluster $MeC(CH_2)_3Sb_3$ displaces the weakly bound THF ligand from $Cr(CO)_5(THF)$ to give $\{MeC(CH_2)_3Sb_3\}Cr(CO)_5$ (Fig. 6.17).

In Section 6.6 the displacement of the π-organic ligand $1,3,5\text{-}Me_3C_6H_3$ from $(\eta^6\text{-}1,3,5\text{-}Me_3C_6H_3)Cr(CO)_3$ by Zintl ions was described. When $(\eta^6\text{-}1,3,5\text{-}Me_3C_6H_3)Cr(CO)_3$ reacts with $[As_7]^{3-}$, the organic ligand is displaced and the addition of the arsenide ligand to the chromium atom is accompanied by the cleavage of one As–As bond.

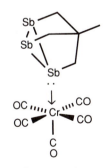

Fig. 6.17 Structure of $\{MeC(CH_2)_3Sb_3\}Cr(CO)_5$.

6.9   Adamantane-type and related clusters

P_4O_6 and P_4O_{10}

Refer to Figs. 3.36 and 3.37 for the structures of the adamantane-like clusters detailed in this section.

Typical reactions of phosphorus(III) oxide, P_4O_6, are summarized in Fig. 6.18. Hydrolysis and halogenation reactions destroy the cluster; reactions with Cl_2 and Br_2 are violent and give the corresponding phosphoryl halides. Oxidation of P_4O_6 occurs with dioxygen or sulfur. The P_4O_6 cluster can act as a Lewis base as shown in the reaction with $Ni(CO)_4$. A carbonyl ligand is displaced by a phosphorus donor at each of four nickel centres giving the tetrametallated complex $P_4O_6\{Ni(CO)_3\}_4$ in which the P_4O_6-cage is retained as the central cluster unit; compare this with the reaction of P_4 in Fig. 2.16a.

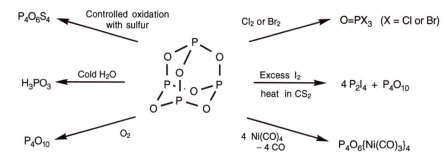

Fig. 6.18 Reactions of P_4O_6.

Phosphorus(V) oxide, P_4O_{10}, reacts vigorously with water giving H_3PO_4 and can be used as a dehydrating agent. Similar reactions occur with alcohols (Eqn 6.37), but if the conditions are not controlled, the alcohol may be dehydrated (e.g. C_2H_5OH to C_2H_4). P_4O_{10} is able to dehydrate a range of compounds, e.g. amides to nitriles, and sulfuric acid to sulfur trioxide. The P_4O_{10} cluster is degraded during the reaction with a mixture of Cl_2 and PCl_3 (Eqn 6.38) and on treatment with NH_3. Ammonium salts of amido-polyphosphates result from the latter reaction and find application in removing Ca^{2+} ions from hard water.

$$P_4O_{10} + 6\ ROH \longrightarrow 2\ P(OH)_2(OR)(O) + 2\ P(OH)(OR)_2(O) \qquad \textbf{Eqn 6.37}$$

$$P_4O_{10} + 4\ PCl_3 + 4\ Cl_2 \longrightarrow 2\ O{=}PCl_2{-}O{-}PCl_2{=}O + 4\ O{=}PCl_3 \qquad \textbf{Eqn 6.38}$$

$P_4(NR)_6$ and $As_4(NR)_6$ (R = Me or ^nPr)

Trimethylamine oxide is a useful oxidizing agent because it readily releases oxygen:

$$\overset{+}{Me_3}N\text{--}\overset{-}{O} \rightarrow Me_3N + [O]$$

Some of the chemical properties of $P_4(NMe)_6$ resemble those of P_4O_6. Oxidation with O_2 at 170°C leads to a polymeric product; controlled oxidation is achieved by using an amine oxide (Eqn 6.39). Oxidation with elemental sulfur gives $P_4(NMe)_6S_n$ ($n = 1$ or 2 when the reaction is carried out in CS_2 at -20°C, and $n = 3$ or 4 when three or four moles of elemental sulfur are used and the solvent is ethanol). $P_4(NMe)_6$ acts as a Lewis base in reactions such as those shown in Eqns 6.40 and 6.41; $As_4(NMe)_6$ mimics $P_4(NMe)_6$ in its reaction with $Ni(CO)_4$. Evidence for the retention of the adamantane-like $P_4(NMe)_6$-core in the metallated products comes from the fact that Lewis bases such as PPh_3 which are stronger than $E_4(NMe)_6$ ($E = P$ or As) displace the cluster from $E_4(NMe)_6\{Ni(CO)_3\}_4$ leaving the $E_4(NMe)_6$ cluster unperturbed.

$$P_4(NMe)_6 + 4\,Me_3NO \longrightarrow P_4(NMe)_6O_4 + 4\,Me_3N \qquad \text{Eqn 6.39}$$

$$P_4(NMe)_6 + 4\,Ni(CO)_4 \xrightarrow{-4\,CO} P_4(NMe)_6\{Ni(CO)_3\}_4 \qquad \text{Eqn 6.40}$$

$$P_4(NMe)_6 + \text{excess}\ B_2H_6 \longrightarrow P_4(NMe)_6(BH_3)_n \quad (n = 1\text{-}4) \qquad \text{Eqn 6.41}$$

The clusters $P_4(NMe)_6$ and $As_4(NMe)_6$ react with methyl iodide but in quite different ways. The phosphorus cluster forms $[MeP_4(NMe)_6]^+I^-$ or, with excess MeI, is cleaved to give $[MeP(NMe_2)_3]^+I^-$. The arsenic cluster reacts according to Fig. 6.19 eliminating methylamine and forming an open cluster. $As_4(NMe)_6$ reacts with HX ($X = F$, Cl, CF_3SO_3) in a similar manner to give $As_4(NMe)_5X_2$. In each case the process can be reversed.

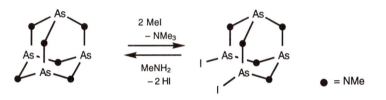

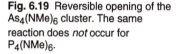

Fig. 6.19 Reversible opening of the $As_4(NMe)_6$ cluster. The same reaction does *not* occur for $P_4(NMe)_6$.

Phosphorus sulfides, P_4S_n and related selenides

Not all the clusters in the family P_4S_n have been equally well investigated. Solid P_4S_3 is resistant to aerial oxidation at room temperature but when dissolved in CS_2, it reacts with dioxygen to give $P_4S_3O_4$. P_4S_3 is very stable in water but is decomposed by alcohols. It hydrolyses slowly in acidic solution. Reactions with elemental sulfur provide routes to higher nuclearity P_4S_n clusters and the reaction with iodine gives a method of preparing P_4S_7 (Eqn 6.42). Zinc metal reduces P_4S_4 causing extrusion of the sulfur atoms and contraction of the cluster framework (Eqn 6.43).

Some interconversions between clusters of the type P_4S_n are shown in Fig. 5.22.

A commercial use of P_4S_3 is in *strike-anywhere* matches. For this purpose it is combined with $KClO_3$; the two compounds detonate when subjected to friction.

$$7\,P_4S_3 + 24\,I_2 \longrightarrow 3\,P_4S_7 + 16\,PI_3 \qquad \text{Eqn 6.42}$$

$$P_4S_3 + 3\,Zn \xrightarrow{1100^{\circ}C} P_4 + 3\,ZnS \qquad \text{Eqn 6.43}$$

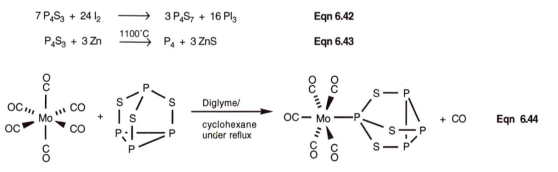

Both P_4S_3 and P_4Se_3 exhibit Lewis base behaviour. In reactions with $\{(Ph_2PCH_2CH_2)_3N\}Ni$, the clusters coordinate to the nickel atom through the unique phosphorus atom to give products resembling $\{(Ph_2PCH_2CH_2)_3N\}Ni(\sigma\text{-}P_4)$ (Fig. 2.16a). P_4S_3 displaces a carbonyl ligand from $Mo(CO)_6$ (Eqn. 6.44). In both examples, the integrity of the cluster is preserved on coordination. In other cases, the cluster is cleaved and is a source of a P_3-ligand (Eqn 6.45) or a P_2S-ligand.

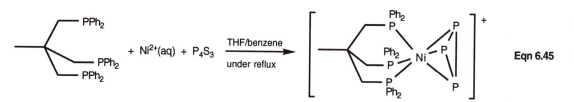

Eqn 6.45

The cluster P_4S_{10} releases H_2S on exposure to air and it hydrolyses in water and in dilute acids. It reacts with NaOH to give $Na_3PS_2O_2$ and H_2S. Below 100°C, P_4S_{10} reacts with alcohols ROH to yield $P(OR)_2(SH)(S)$ and H_2S; the rate of reaction depends on the size of the substituent R. At higher temperatures, the products are $P(OR)_3(S)$ and H_2S.

Sulfur atoms can be abstracted from P_4S_{10} by treating it with PPh_3, PCl_3, or PBr_3 (Eqn 6.46) and this selective degradation is used synthetically (see Fig. 5.22). P_4S_{10} is completely degraded when it is heated with PCl_5 or SF_4 (Eqns 6.47 and 6.48).

$$P_4S_{10} + Ph_3P \rightarrow P_4S_9 + Ph_3P{=}S$$

Eqn 6.46

$$P_4S_{10} + 6\,PCl_5 \rightarrow 10\,Cl_3P{=}S \qquad \textbf{Eqn 6.47}$$

$$P_4S_{10} + 5\,SF_4 \rightarrow 4\,PF_5 + 15\,S \qquad \textbf{Eqn 6.48}$$

6.10 Tetrasulfur tetranitride (cyclotetrathiazene) and tetraselenium tetranitride (cyclotetraselenazene)

Selected chemical reactions of tetrasulfur tetranitride S_4N_4, are given in Fig. 6.21. The cluster opens into a ring during reduction, fluorination to $S_4N_4F_4$, controlled chlorination to $S_4N_4Cl_2$, or reactions with Lewis acids. Fluorination occurs at the sulfur atoms leading to a sulfur–nitrogen ring for which a localized bonding scheme is appropriate (Fig. 6.20). Depending on conditions, fluorination of S_4N_4 can cause cluster degradation; excess AgF_2 leads to $N{\equiv}SF$ and $N{\equiv}SF_3$, and reaction with HgF_2 gives $N{\equiv}SF$. Reduction to $S_4N_4H_4$ (Fig. 6.20) occurs with N–H bond formation. Electrochemical reduction gives the radical anion $[S_4N_4]^-$ but this is unstable above 0°C. In a sealed tube, S_4N_4 reacts with liquid Br_2 to give $[S_4N_3^+][Br_3^-]$. Similar reactions occur with liquid ICl and IBr.

$[Br_3]^-$

Fig. 6.20 Structures of $S_4N_4H_4$ and $S_4N_4F_4$.

Solid S_4N_4 reacts with Br_2 or ICl vapour to yield conducting polymers (Fig. 6.21). Lewis acids may function as simple acceptors (e.g. BCl_3 or H^+)

to give adducts (e.g. $S_4N_4.BCl_3$ or $[S_4N_4H]^+$) in which a lone pair of electrons is donated from one nitrogen atom of S_4N_4 to the Lewis acid. If the Lewis acid is an oxidizing agent (e.g. AsF_5, SbF_5, $SbCl_5$) the S_4N_4 cluster undergoes a two electron oxidation (Eqn 6.49).

$$\left. \begin{aligned} S_4N_4 &\rightarrow [S_4N_4]^{2+} + 2\,e^- \\ AsF_5 + 2\,e^- &\rightarrow AsF_3 + 2\,F^- \\ AsF_5 + F^- &\rightarrow [AsF_6]^- \end{aligned} \right\} \quad S_4N_4 + 3\,AsF_5 \longrightarrow [S_4N_4][AsF_6]_2 + AsF_3 \qquad \textbf{Eqn 6.49}$$

When S_4N_4 vapour is passed over hot silver wool, S_2N_2 forms. Passage over quartz wool leads to $(SN)_x$. This polymer exhibits one-dimensional electrical conductivity and becomes superconducting at 0.26 K. Cleavage of the S_4N_4 cluster also occurs during reactions with nucleophiles; attack by Ph_3P leads to $Ph_3P=NS_3N_3$.

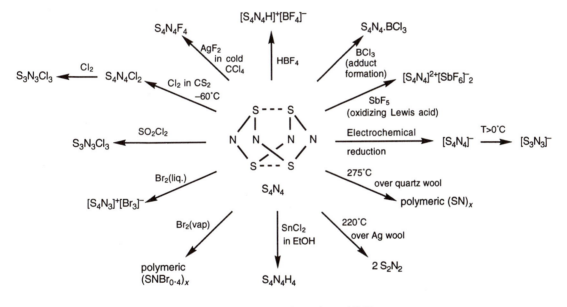

Fig. 6.21   Selected reactions of S_4N_4.

Like S_4N_4, Se_4N_4 is potentially explosive. Its chemical properties have not been as fully investigated as those of S_4N_4. Pyrolysis causes degradation to selenium and dinitrogen. With bromine, the cluster is cleaved to give $[NH_4]_2[SeBr_6]$ and reduction with hydrazine yields selenium and ammonia.